Freshwater Fish Pond Culture and Management

Marilyn Chakroff
Peace Corps

Illustrated by

Joan Koster

Edited by

Laurel Druben

United Book Prints

1st Reprint in India, 2015

Published by:

United Book Prints
4806/24, Ansari Road,
Daryaganj,
New Delhi - 110 002
Tel.: 011-41511055

ISBN: 978-93-83692-04-0

Print: 2021

Edition: 1977, 1978, 1981 & 1982

Printed in India

Philippines for three years. in a number of fisheries programs. Ms. Chakroff, who holds a B.S. in Biology, now is an advanced deqree candidate in the field of Environmental Communication at the State University of New York, in Syracuse. This manual is written out of her first-hand experiences as a Peace Corps Volunteer.

Joan Koster, the illustrator, has been a VITA Volunteer for more than 3 years. She is a teacher, professional artist, and writer. Ms. Koster, who has travelled and studied in Greece for a number of years, currently is preparing a manuscript on looms and weaving.

OTIHER CONTRIBUTORS

Many thanks are due here to a number of people who aided the preparation of this manual:

Dr. David Hanselman, Dr. Peter Black, and Dr. Robert Werner, -- Faculty of the College of Environmental Science and Forestry, State University of New York, Syracuse.

Dr. Shirley Crawford, Agricultural and Technical College, State University of New York, Morrisville.

William McLarney, New Alchemy Institute, Woods Hole, Massachusetts.

A.F. D'Mello, Hawkesbury Agricultural College, New South Wales, Australia.

Richard T. Carruthers, Bioproducts, Inc., Warrenton, Oregon.

Dr. William Ribelin, Department of Veterinary Science, University of Wisconsin, Madison.

A special note of thanks is due John Goodell, VITA, for his layout work and staff assistance with this manual.

Table of Contents

1 Introduction

What is Fish Culture?

Fish culture is the growing of fish in ponds. Growing fish in ponds, from which they cannot escape, allows feeding, breeding, growing, and harvesting the fish in a well-planned way.

Fish culture is one form of aquaculture. Aquaculture is the science which deals with methods of growing (cultivating) animal and vegetable life in water. Some other kinds of aquaculture are concerned with growing frogs, oysters, seaweed, and even rice.

History of Fish Culture in Ponds

Growing fish in ponds is a very old practice. Carp were cultured as long ago as 2698 B.C. in China, where they were grown in ponds on silkworm farms. Fish culture seemed to occur whenever civilization was settled for a long period of time. For example, fish culture was done in ancient Egypt and in China, which has had a continuous civilization for over 4,000 years. The first written account of fish culture in ponds was by Fan Lai, a Chinese fish farmer, in 475 B.C.

The ancient Romans introduced carp from Asia into Greece and Italy. By the seventeenth century (1600's), carp culture was being done all over Europe. A book written in England in 1600 by John Taverner gives the details of good pond management and talks about growing the common carp. Taverner also wrote about pond construction, fertilization and feeding. Another book, written in 1865, gave the details of the stripping methods of spawning fish. The methods of culturing common carp have not changed very much since that time.

The common carp is still a very important pond fish. In addition, today, other fish also are being cultured in ponds. Some of the most well-known are fish of the tilapia genus, like *Tilapia nilotica* and *Tilapia mossambica*. Some of the other Chinese carps -- the silver, grass, and big head carps -- also are often used in pond culture. Most importantly, countries all over the world are using time and money to discover which of the fish commonly found in their own waters will grow well in fish ponds.

Why Fish are Grown in Ponds

The practice of culturing fish in ponds developed because growing fish in ponds is a more useful practice, for some purposes, than trying to catch fish from lakes, rivers, or streams. For example:

- Many interested people discover that building a fish pond close to home is possible and far more convenient than going to the nearest market or river. Ponds can be built wherever the soil, shape of the land, and water supply are right. This may sound as if a lot of factors are involved. But since a wide variety of soils, land shapes, and water supplies can be used for pond culture, a fish pond can even be made from a rice paddy or an unused grain field.

- It is easier to get fish out of a pond than it is to catch a fish from a river or stream. Also, the number of fish taken out of a pond can be controlled. But it is very difficult to know how many fish can be caught in a river or stream or lake at any one time. When the farmer goes to his fish pond to get dinner, he knows he can take out the number of fish he needs --quickly and easily.
- Fish growth can be controlled. The fish can be fed extra food to make then better for market; natural enemies can be kept from killing the fish. For a person who relies on fish for his food or his income, these are important factors.
- The only fish grown in a pond are the ones the farmer wants to grow. When he takes a fish out of his-pond, the farmer knows what kind or kinds he will be getting. When he catches fish in a lake, stream, or river, many of the fish will not be the ones that are good to eat or to sell.
- Growing fish in ponds allows the farmer, or other fish grower, to produce fish cheaply, and to have a supply of fish available on his own land. Fish in ponds belong to the pond owners; fish in the rivers and lakes do not.

Why Growing Fish is Important

There are some very good reasons why a farmer or small land owner might be interested in fish farming:

- Fish are an important food source.
- Fish farming can help a farmer make the best use of his land.
- Fish farming can provide extra income.

There may be additional reasons; you and the pond owners can determine these from the local situation. The three points listed above are very broad, however, and apply, at least in part, to most situations. Therefore, each point is discussed more fully below.

FISH AS FOOD Farmers know that all living things need food, and that without food, living things die. However, they are not as likely to know the characteristics of food which make it valuable (or not) to the body.

Food is important because it provides proteins, vitamins, minerals, fats, and carbohydrates. These things are called nutrients: they are materials that the body must have to live and grow. Every kind of food has different amounts of each of these nutrients. For example, some foods contain more protein; others have more fat than protein.

Because foods contain different amounts of proteins, fats, and carbohydrates, for example, it is necessary to eat a number of different kinds of food to get the right amounts of each nutrient. All the foods together then give the body what It needs to grow.

The food that people eat is called their diet. Eating the right kinds of food --foods that give the body the right amounts of proteins, fats, etc. --is called eating a balanced diet. People who eat a balanced diet usually are healthy and strong; people who do not eat the right kinds of food are more likely to be weak and get sick.

Proteins are the most important part of food. Protein is made of carbon, hydrogen, and nitrogen. These are called elements. The combinations of elements in protein make it the most useful nutrient. Foods that contain a lot of protein are especially good for people to eat. And fish contains a lot of protein.

The table on the opposite page shows a list of foods that humans eat. The first number beside the food shows the number of grams of protein in the food when it is fresh. The second number tells how many grams of protein there are in food which has been dried. The table shows that fish --whether fresh or dried -- is a very good source of protein. (100 gm of dried fish contains more protein than 100 gm of fresh fish only because dried foods have water taken out. Therefore, 100 gm of fresh fish weighs less when it is dried).

If the farmers in your area already eat a lot of fish, or like fish, fish farming for food may not be hard to introduce and have accepted. If they do not eat fish often, you will have to keep this in mind when you talk about fish as a healthy food. Food just may not be the most important reason, from their point of view, for wanting to grow fish.

PROTEIN CONTENT OF FOODS*

Food	Fresh, gms protein per 100 gm	Dried, gms protein per 100 gm
FISH		
Fatty (herring)	17	46
Non-fatty (haddock)	16	84
MEAT		
Beef	20	67
Pork, loin	20	67
Liver	67	67
DAIRY PRODUCTS		
Milk	3.4	26
Eggs	12	46
CEREALS		
Wheat	12	14
Maize	10	11
Oats	10	11
Rice	8	9
OIL SEEDS		
Soya	33	37
Cottonseed	20	21
Sesame	21	22
GREEN LEAFY VEGETABLES		
Cabbage	1.4-3.3	24
Spinach	2.3-5.5	26
ROOTS		
Cassava (manioc)	0.7	2

Potatoes	2.1	9
Yams	2.1	7
Plantains	1.0	3

*These values are estimates only; the amount of protein varies according to the age, size, and quality of the food, and how it was cooked and stored.

Source: Aylward and Jul (1975)

But there are other reasons you can offer a farmer. For example, a farmer may consider cultivating fish if he realizes that fish are easy to grow, cheaper than some kinds of meat, available as food all year round, etc. You will have to see which combination of arguments works best for getting farmers interest.

BETTER LAND USE Some farmers may be more interested in fish farming when they realize they can accomplish two purposes: provide a reliable food supply and make the best possible use of their land.

"Fish farming" is a good thing to call "fish culture" because it can start the farmer thinking about raising fish with the same kind of planning and land-use management ideas that he puts into raising crops.

Whether the farmer raises fish, crops, or animals, he is using his land in certain ways. His aim in all cases is to increase the production of food and the yield from the land. What farmers, and other people, often do not realize is that fish culture can help get more out of the land. Here are a few ways in which fish culture can help support and extend a farmer's land use:

- Land gets tired when it is used for growing the same crop year after year. These crops use up nutrients in soil, and they begin to grow poorly. Fish ponds can be built on this land and fertilized to provide food for the fish. After a few years of fertilizing and growing fish, the soil inside the pond regains some of the nutrients used up by the growing of crops year after year. The land can then be used for crops again.

- Some farmers own land that may not be very good for growing crops: it is too sandy, for example. But there are ways of building fish ponds in sandy soil. So the farmer would be able to use land that was once not of much value to him.
- There are many ways that fish farming can fit into the farmer's plan for his land. The important thing is that all of these ways help the farmer make the best use and get more out of what he has --readily, and often without much expense. For example, a farmer who grows paddy rice can grow fish in that paddy; fish ponds can be built as part of water supply and irrigation systems; vegetable scraps and animal manures can be collected and used for fertilizing ponds. The farmer should know that a farm with a fish pond or ponds can give a total food yield that is higher than a farm with no fish ponds.

The following diagram illustrates some of the ways in which the fishpond fits into the farm: The same water source is used by both the garden and the fish pond; the mud from the bottom of the pond makes good fertilizer for the garden; vegetable matter from the garden can be used to fertilize fish ponds; manure from the animals can be used for the pond and parts of fish can be used to feed animals; etc.

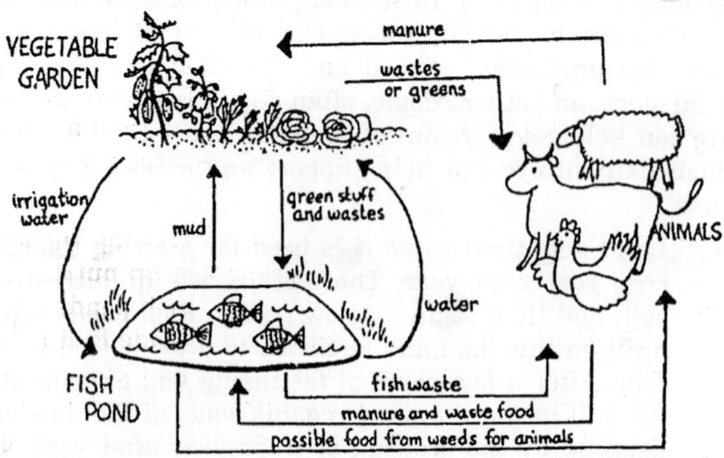

ADDED INCOME Fish ponds can be quite small, or they can be large. They can be made using expensive equipment and drainage systems, or they can be dug using hand tools and drained by a bamboo pipe. Fish can grow successfully in both of these types of pond, as long as the ponds are managed correctly.

If the major reason for building the fish pond is to get increased and better food for his family, a farmer certainly does not need fancy ponds or expensive equipment. Fish ponds can be very inexpensive to keep. Fish do not require fancy foods. Many ponds provide all the food the fish need. But besides the foods they find in the water itself, some fish eat leafy garbage, mill sweepings, beer residues, spoiled grains, broken rice, and many other waste products that might not otherwise be used.

A farmer makes his income go further by growing more of the family's food and by selling left over fish the family cannot eat.

Growing fish to sell can also be very profitable. But the costs involved in getting started and in maintaining the effort are greater: if the farming is to be a solid. commercial enterprise, then more ponds, more time, more money, and nearby market places are needed. The business may or may not show a profit right away; in fact, the chances are that it will not. A farmer might be better advised to start small and work into a bigger enterprise slowly as he learns to manage the art of growing fish in ponds.

A Word about Cooperation

Often fish ponds are built by cooperatives. A cooperative is an organization of people in an area who come together to do something they could not or would not do alone. In this way, four or five people or families can pool their resources and build a fish pond operation together. Sometimes an entire village will form a cooperative and will build and operate a pond as a group. This kind of cooperation makes possible better pond construction and management. A fish pond cooperative may be a good way for a village to improve the diet of the community and to sell enough fish to maintain the enterprise. If the farmers in your area are not interested in, or are concerned about, building ponds individually, a cooperative may be a very acceptable idea.

Getting Ready to Plan a Fish Farm

A farmer or other person interested in growing fish should read the following list carefully before going further. The following factors must be considered before the farmer builds his fish pond. Many pond owners have small fish ponds that are only used for their own families, but a farmer who sells fish must look for a market and a way to get his fish to that market. It does no good to harvest fish which cannot be sold or used by the farmer and his family.

- Is the soil able to hold water for a fish pond?
- Is there an adequate supply of water for a pond?
- Is the land a good shape for a fish pond?
- Is the pond area close to your home?
- Who owns the land where the pond will be built?

- Are there enough people to help build and harvest the pond?
- Can the equipment for building a pond be built, borrowed, or bought?
- is there a marketplace nearby?
- Are there roads from the pond area to a market place?
- Are the roads passable even in the rainy season?
- Is there a good way to get the fish to market?
- Is there a vehicle available for transportation, if necessary?
- If there is no market nearby, or if it is hard to get to the market, can the fish be kept by drying, smoking, or salting?
- Is there enough food for the pond fish?
- Are there fertilizers available?
- Do the people in the area like fish? Do they eat freshwater fish?
- Can the people in the area afford to buy the fish produced in the pond?

If the farmer can answer yes to the questions which most fit his situation, he has a good chance of having a successful fish pond. But he must consider these factors. Each is discussed in detail in the "Planning" sections.

2 Planning : The Site and the Type of Fish Farm

Before construction can begin, the farmer must look over his land to choose the place or places where ponds can be built, and decide what kind and how many to build. He must also decide on the kind of fish culture he wants to do, and on the type of fish that he wants to raise. He must look at his resources and his needs very carefully before he actually begins building and operating a fish pond. This section will give information to guide the farmer in the planning of ponds and kind of fish culture.

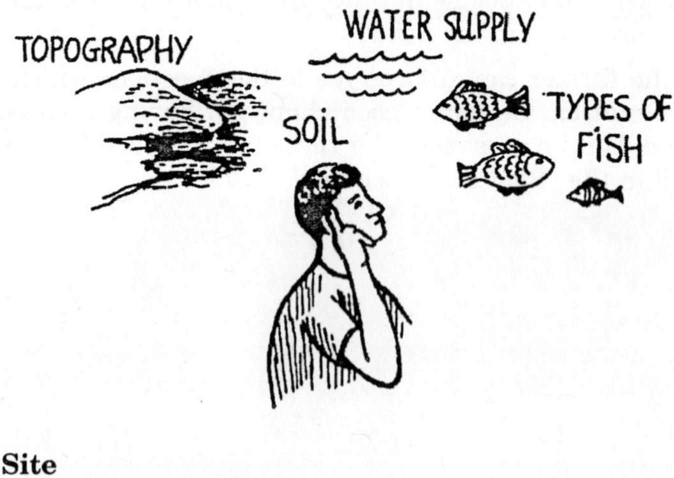

The Site

One of the most important parts of planning is finding the right place (selecting the site) for the pond. Fish ponds use the land in a different way from agricultural crops such as rice or wheat, but fish also are a crop. And when a farmer builds a fish pond, he is choosing one use of his land instead of some other use. If the site for the pond is well-chosen, the pond can be more productive than the land by itself. But if it is not chosen well, the farmer may lose, or, at best, gain nothing from his fish pond. When considering a site for the fish pond, the farmer should remember and consider several points that were made in the introduction:

- Often poor agricultural land can be turned into very good fish ponds. In general, the better the soil of an area, the better the fish pond. But this does not mean that a pond cannot be built on poor land. It does mean that the farmer will have to work harder to maintain the pond and the fish.
- If the pond is built on agricultural land which is not producing good crops, but the pond is cared for well, eventually the pond bottom soil will become more fertile than It was before. If this pond is a large one, after harvesting the fish, the pond can be planted again with a land crop, like corn, and allowed to grow. Then when the corn is harvested, the land can be turned back into a fish pond. This means that a farmer can get two good uses out of his land instead of one poor crop.
- Other farmers may want to grow fish in rice paddies by digging trenches around the edges of the paddy for fish to swim in. This is another way of culturing fish which will be discussed in somewhat more detail later in the manual.

The point of the discussion above, is that a fish pond is just one use that a farmer's fields can have, and the choice of how the land can be used is important.

There are three factors that work together to make a good site for a fish pond:

- Water supply
- Soil
- Topography

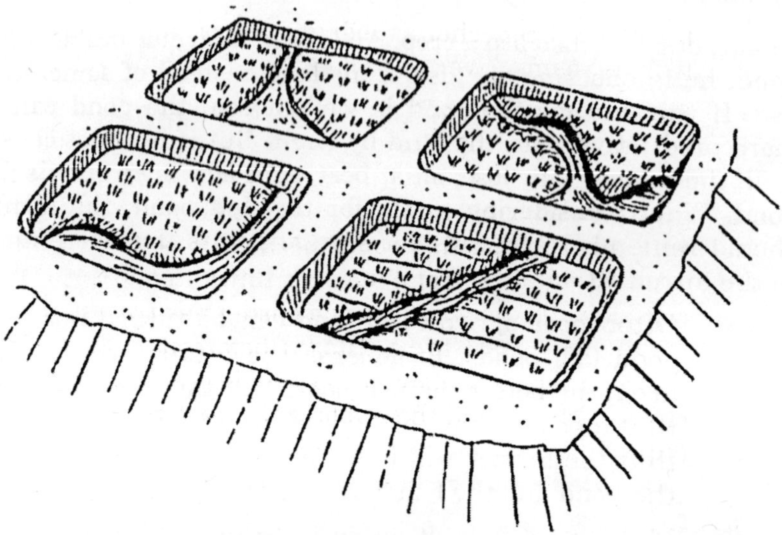

Four types of ditches dug in paddy field for fish

WATER SUPPLY Water supply, soil, and topography all are important, but water supply is the most important factor in selecting a site. Fish depend upon water for all their needs: fish need water in which to breathe, to eat, and to grow and reproduce. If a site has water available year-round, that site meets its first test easily. If water is not available all the time but there is some way to store water -- in large tanks, barrels or drums, in depressions, ponds, or wells -- for use when the natural water supply is low, then that site may still be all right. The key, of course, is that water must be available at all times and in good supply.

<u>Where Can Water for Fish Ponds Come From?</u> Water used in ponds comes from many sources:

- *Rainfall.* Some ponds, called "sky" ponds, rely only on rainfall to fill their need for water.
- *Run-off.* Some ponds are gravel and sand pits which fill when water from the surrounding land area runs into them.
- *Natural waters.* Most ponds are filled with water that comes from natural springs or wells, or with water

that has been channelled (diverted) and brought in from streams, rivers, or lakes.

- *Springs.* Some ponds are built where there is a spring to supply water. Spring water is water under the ground that has found a way to get out. It leaves the ground and becomes a stream as it flows away. Spring water is good for fish ponds because it is usually clean (uncontaminated) and has no unwanted fish or fish eggs in it. If the water from a spring has travelled very far, It may need to be filtered before it is used for a fish pond. But filtering is easy to do (see the "Construction" section) and the important fact is that the water supply is available.

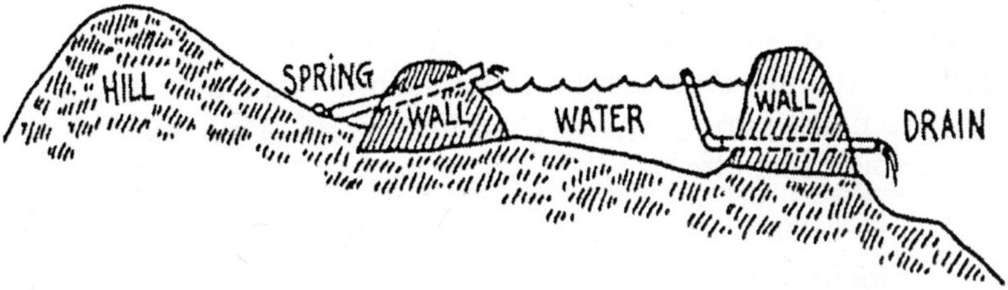

Pond built using a spring as a water source

- *Wells.* The best source of water for a fish pond is well water. Well water has few contaminants and, if the well is a good one, the water is continuously available. Well water and spring water, however, are both often low in oxygen content. Fish need to have oxygen in their water to live. Since this problem is over come easily (see water quality information in the section on "Preparing the Pond") the major factor to be considered here is an adequate water supply.

Most fish ponds use water that comes from a stream, river, or lake. A diversion ditch or channel is dug between the water source and the pond to take water from source to pond. This is a good way to fill a pond because the water can be controlled easily. When the pond is full, the channel can be blocked with a gate or a plug (see "Construction" section), and the water will stop moving into the pond.

There can be problems with this kind of water supply; for example, often in tropical areas streams flood in the rainy season. This extra water can be dangerous to the pond and must be diverted away from the pond by a channel built for that purpose. IT IS BEST NOT TO CHOOSE A PLACE THAT IS KNOWN TO FLOOD WHEN CHOOSING A WATER SUPPLY AND SITE FOR A POND. When a pond floods, all the fish escape and the pond is empty at harvest time.

If the water for the pond is being taken from a stream, lake, or river, then the farmer should plan to filter the water carefully when filling the pond. Water from these sources sometimes contains unwanted fish or fish eggs. Filtering prevents these fish or eggs, and other harmful animals, from entering the pond.

Quality of the Water Supply. Finding an adequate water supply is the first step. Then the farmer has to check that supply to make sure it can be used for a pond. This check of the water should include:

- looking at the water, smelling it and tasting It.
- looking to see if there is a family upstream who take baths in the water before it gets to the pond.
- making sure that there is no family or village downstream that depends upon the source for their drinking water.

If the water supply seems all right, the farmer must also find the answers to some other questions. Where the water comes from, how far it travels to get to the site for the pond, and what kind of soil it travels over will all affect the quality of the water. These questions and their answers tell what must be done to make the water right for a pond:

- Is the water very clear? Then the farmer may have to fertilize the pond because there are not enough nutrients in the water.
- Is the water very muddy? Then it will have to settle before it is used in the pond: a special place will have to be made where the mud can settle out of the water before the water goes into the pond.

- Is the water a bright green? It probably has a lot of fish food in it.
- Is the water a dark, smelly brown? It may have acid in it, and the farmer will have to add lime to the water.

There are many things which can be done to make water good for a pond. If the farmer knows his supply and the kind of water he has, he can take the steps necessary to use his supply well.

SOIL The second important part of site selection is the soil of the area. The soil of the pond must he able to hold water. It also contributes to the fertility of the water because of the nutrients it contains.

Ability of Soil to Hold Water. The best soil for a pond contains a lot of clay. Clay soil holds water well. When a place with a good water supply is found, the farmer must test the soil. He can tell a lot about the soil simply by feeling it. If the soil feels gritty or rough to the touch, it probably contains a lot of sand. If it feels smooth and slippery, it probably means there is a lot of clay init. This smooth soil is good for a fish pond.

A very good way to tell if the soil is right for a fish pond is to wet a handful of soil with just enough water to make it damp.

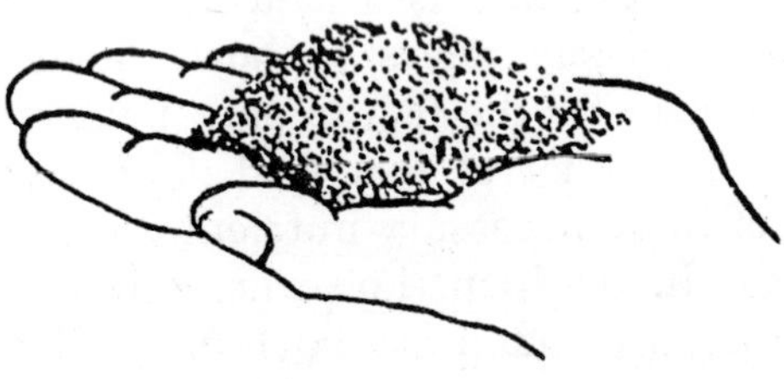

Then squeeze the soil.

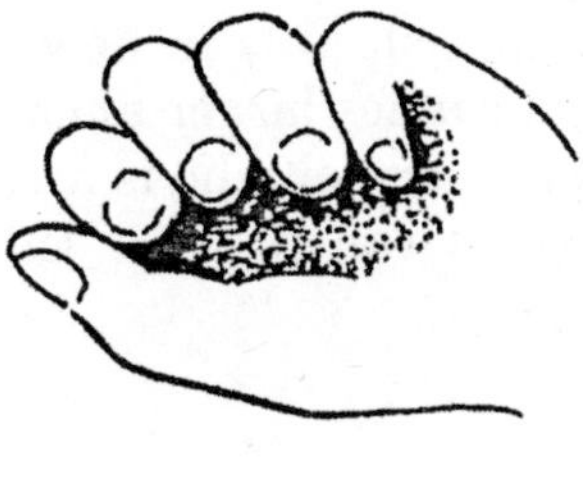

If it holds its shape when the farmer opens his hand, it will be good for a pond. Remember, the more clay in the soil, the better it is for building a pond.

If the soil is sandy, or does not contain much clay, the farmer can still build a pond. There are ways of building ponds in these soils. But he should be aware that building a fish pond in such soils requires more effort and may not be as successful. Digging test holes will tell the farmer what his soil is.

Larger ponds can be built in soils with clay. If the soil is rocky or has shifting sand, etc., only small ponds are possible. If there are other locations available, the farmer would be wise to see if there is another place with soil better suited to the fish pond. More information on soil is included in the "Construction" section.

Ability of Soil to Provide Nutrients. Soil also contributes to the pond's fertility. Fertility is a measure of the nutrients in the pond, and it simply refers to how much food there is available in the pond for the fish to eat. A very fertile pond is one which contains a lot of fish food. The soil of the pond contains some of these necessary nutrients --like iron, calcium, and magnesium. In addition, however, soil also can contain acids; these substances often are harmful to fish. Whatever a soil has in it is drawn into the pond by the water and thus comes in contact with the fish. Sometimes after a heavy rainstorm, there are big fish kills in new ponds. This happens because the heavy rain carries larger amounts of acids from the soil into the pond. So the farmer who is aware of the kind of soil he has for his fish pond can prevent this problem before it happens.

REMEMBER: One good Indicator of the quality of soil Is whether it has been used for growing crops. If crops grow well in that location, the soil will probably be good for the fish pond. If crops did grow well there before the nutrients were used up, then it will probably still be free of harmful substances.

TOPOGRAPHY The third factor in site selection is topography. Topography is a word used to describe the shape of the land -- whether it is flat or hilly, upland or lowland, etc. The topography of the land determines the kinds of ponds which can be built. Ponds can be built in valleys or on flat ground. They can be square or rectangular, or uneven in shape. They can be large or small. All of this is determined by topography of the land, as well as by the farmer's requirements.

The most useful topography for fish ponds is that which allows the farmer to fill and drain ponds using gravity. Ponds built on a slope, for example, can be drained easily. If ponds are located on flat land, the pond must be built with a slope inside it so it can be drained by gravity, or it will have to be drained using a pump.

Slope. If the farmer looks at a hillside, he can see that It rises. It is higher at one point than at another. This difference in height, from high to low point, is the slope of the land. In more scientific terms, slope is the relationship between the horizontal distance (length) and the vertical distance (elevation) over a piece of land.

Slope is usually written as a ratio (1:2) or as a percentage (5%). A slope of 1:2 means that for every change in length of 2 meters, there Is a change of 1 meter in height. A slope of 5% means that for every change in length of, say, 100cm, there is a change in height of 5cm. Pond bottoms usually have a slope of 2-5%, whether they are on level ground or in a hilly area. As long as the pond bottom has a slope, it can be drained completely.

A farmer does not require a scientific understanding of slope to build a pond. He does need to know how the shape of his land determines the best place for building ponds. Ponds built in hilly places often are made part of the hill. The picture on top of the next page, of a pond with a spring as a water source, shows how the slope of the land has been used to set up the pond's drainage system.

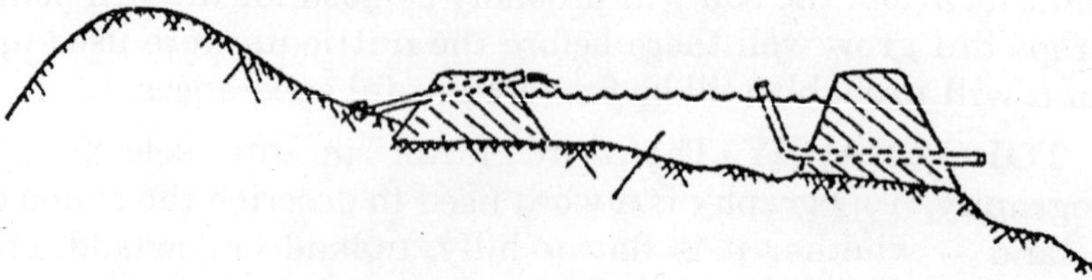

In flatter areas, ponds are usually square or rectangular because it is easier to use a harvesting net in ponds of these shapes.

The farmer will learn quickly to recognize by sight the slope that is best for a pond. Because a slope is so important, the first thing a farmer should look for Is a site with a slope and a water supply. If he can use a natural slope for his pond, the pond will be cheaper and easier to construct.

The best places to look for such combinations of slope and water supply are where water collects from streams and flows through the valley at the bottom of a slope. If the pond is built on the slope above the water flow, water drained from the pond can flow directly into the stream. Water might be brought to the pond in a number of ways depending upon the situation -- by streams running down the slope upon which the pond is situated, for example. Another good place to look for a good combination of slope and water supply is on plains or flattish ground between hills. These plains often receive water from brooks or streams.

There are many possibilities. The important thing is that the farmer look for a topography that makes fish farming as easy and as successful as possible.

The Type of Fish Farm

After the farmer has found a site or sites for his fish pond, he must consider what kinds of fish culture are possible in the space he has available. He also must decide what his resources will allow him to get started. This planning is necessary because the answers will determine the number of fish ponds the farmer builds and the kind of fish he will want to culture. The following pages present a range of ideas concerning the kinds of fish farm operations (raising fish or breeding fish); the types of pond used in fish culture; fish culture in one or several ponds; advantages of small and large ponds; and mixing or separating fish types and sexes. A discussion of these subjects will provide the farmer with the background he needs to decide what kind of fish farm is possible for him, given his resources and the kind of fish he wants to raise.

A NOTE OF CAUTION Before a farmer even begins, however, it is important for him to include in his planning the fact that some fish will die. This is an extremely important fact for the first-time fish grower to understand. It is very natural for some fish, the weaker fish, to die in ponds. As long as fish are protected in ponds and are well taken care of, fewer fish will die in ponds than would die in natural waters. But a farmer who does not expect some death may get discouraged and give up before he has given his pond a chance to work. It is never too early to introduce this idea.

KINDS OF FISH FARM OPERATION In nature, many fish never reach adult size because they are eaten by other animals (predators), or they die from disease or lack of oxygen. In fish culture, the farmer tries to control the pond situation in order to produce more fish. In ponds, predators and so on can be controlled so that the pond yields more fish per hectare than do natural waters.

There are two major kinds of fish farms --those which breed fish and raise the fry, and those which rear fry and fingerlings (the young fish) to market size. So the farmer, after finding possible sites, etc., most decide if he is going to breed his fish and raise the fry. Or if he is going to buy fry and fingerlings and rear them to market size, not getting involved in breeding.

Breeding fish requires more time and more ponds than simply rearing fingerlings. And building more ponds can be more expensive and require more ongoing management. So the farmer must finally determine his reason for raising fish: to eat; to sell; to use his land better; or all of these. He will have to have all these things firmly in mind so that he can:

- build the right kinds of pond.
- build the right number of ponds.
- stock the right kinds of fish.

TYPES OF PONDS The types of pond a farmer can build depend on water supply, soil, and topography, the factors which were just discussed. The two types of pond most often built are barrage ponds and diversion ponds. Many aspects of the construction of these ponds are the same. The main difference between these two types of pond is the water source.

Barrage Ponds. These ponds are usually filled by rainfall or by spring water. A spring, for example, sends water flowing through a small valley or down a slope into a low place. Or a spring bubbles from the ground into a natural depression. The pond is formed by collecting water at the base of the valley and in the low places. The farmer does this by building a wall (dam) which holds the water inside what now is the pond area. The wall keeps the water from entering and leaving except as needed.

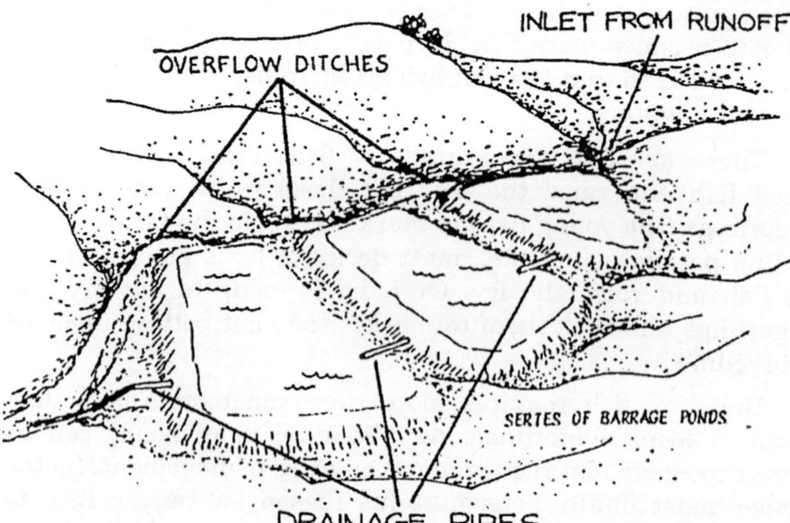

The number of pond walls the farmer must construct depends upon the land and on how he fixes his drainage system. A barrage pond usually needs only one wall --the main wall between the water source and the pond area. One kind of drainage system called a sluice (see "Construction" section) can be used to let water both in and out of the pond. There are also a number of simple drainage systems which can be used that do not require any complicated construction.

Barrage ponds should not be built where the flow of water is too great: it is difficult to keep the water from breaking down the wall if the pressure of the water is too great. Brooks and streams which flow well, but not too strongly, make good sources for barrage ponds.

Even when the flow of water is not great, however, barrage ponds require overflow channels. Because barrage ponds are usually built in low areas, they are likely to fill up in heavy rains. Overflow channels are any kind of system which can be set up to stop the pond from collecting too much water. The overflow takes extra water away from the pond. If this extra water is not taken out, the pond wall may break. Therefore, the overflow system is needed to help the drainage system handle the flow of Water when there is too much water in the pond.

The overflow system can be wide grooves cut into the top of the wall toward the ends away from the middle; it can be large hollow tree trunks which are set into the tops of the wall and work as pipes to drain the water into ditches, or even to carry the water into storage areas for use later when the water supply is low. Another kind of overflow can be ditches, dug into the ground above pond level, which take the extra water away when the water rises to that level.

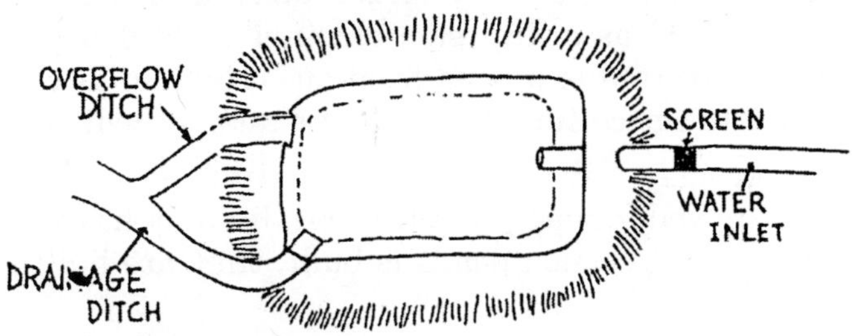

An overflow often is not screened, because if something large catches on it, the pressure of the water behind It might cause the entire wall to break. This fact results in a loss of fish at time of flooding.

Diversion Ponds. These ponds are made by bringing (diverting) water from another source like a stream or river. Channels are dug to carry the water from the water source to the pond.

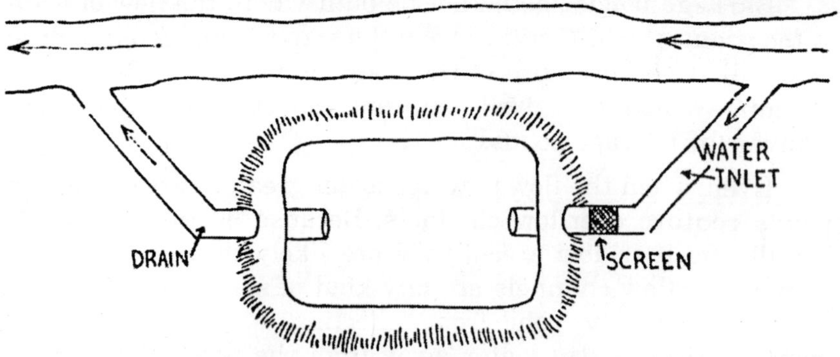

Diversion ponds can be made in a number of ways. Sometimes a pond is dug in flat ground or can be made by slightly enlarging a natural depression in the land.

These ponds, like the barrage ponds, require walls depending upon the topography of the land, the drainage system used, etc. A pond dug in flat ground often requires four walls; a pond built in a natural depression may not.

With a diversion pond, the water is always brought to the pond instead of running directly into the pond. Water can be diverted in a number of ways. For example, a small stream which gets its water from a larger stream nearby can be dammed and used as a diversion channel to feed a pond. Or water can be diverted to a pond from an irrigation ditch which carries water to agricultural crops from a nearby well or lake.

A farmer may have one diversion pond, or if his space allows and the water supply is sufficient, he may have several. When a series of diversion ponds is built, they are built in one of two ways:

- *Rosary system.* These ponds are built one after another in a string. In this system, all the ponds drain into each other and must be managed as if they were one pond. Therefore, if the first pond in the series (the pond with the water inlet) is full of predators which must be poisoned, all the other ponds in the system have to be harvested (have the fish taken out) and drained before the first pond can be poisoned and drained.

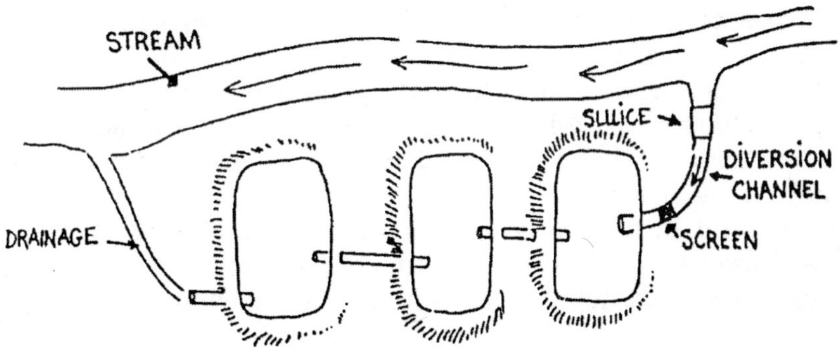

- *Parallel system.* In this series, each pond has its own inlet and outlet. Therefore, each pond can be managed as a separate pond.

Each kind of pond is going to have advantages or disadvantages depending upon the farmer's situation. A parallel system of diversion ponds, in most cases, is a better system. But rosary systems are cheaper and easier to build, and therefore, more possible for some farmers to undertake. Also, if the water source is good, and can be kept free of predators and unwanted fish, and if the management of the pond is done well, a rosary system can be very successful.

Diversion ponds are often better than barrage ponds because they are less likely to overflow, and the water source is often more dependable throughout the year. But barrage ponds require less construction and are likely to be cheaper. In addition, for some farmers, barrage ponds are the best, and perhaps the only, way for them to use their land for fish ponds.

The art of constructing and planning a fish pond or fish operation is very much an individual thing. There are basic ways of using resources, for example, land and water sources. But the exact shape and type of fish pond must be decided by the farmer for his situation. There are many ways of making fish ponds which will work, and the "right" way for any given farmer is the way which works best for him. Many aspects of fish farming are determined by experimenting with pond operation, but much can be done by good planning before fish pond construction.

Therefore, the farmer must look at his sites and consider the types of ponds he can build from the viewpoint of the number, size, and depth of the ponds he is going to need. If, for example, the farmer thinks he has a good area for a diversion pond, but hits solid rock at 1m and needs a pond 2m deep, he can find this out before he invests a great deal of time and money. If he has room for two small diversion ponds and a barrage pond, or for a large diversion pond and a barrage pond, he can base his decision on what kind of pond to build upon the number, size, and depth of pond he needs for what he will be doing.

<u>The Number of Ponds.</u> The number of ponds depends on the possible sites and on what the farmer plans to do with his fish ponds. If he is going to raise fingerlings to market size, he will need one or a few "rearing" ponds. If a farmer plans a larger operation in which he will breed fish for the eggs and fry, he will need space for nursery pond, rearing pond, and a pond for brood stock. Nursery ponds can hold eggs and fry until they are fingerling size; rearing ponds hold the fingerlings until they are market size; brood ponds hold the fish to be used for breeding.

It is possible to breed fish in a corner of a large, single pond, and a farmer interested in raising fish for his own use may want to do this. But a farmer interested in marketing fish probably will want at least two large ponds. If he has two medium-large ponds, he can use one for rearing fingerlings and one for brood stock. Eggs and fry can be taken care of in very small ponds or even containers.

<u>The Size of Ponds.</u> The size of ponds depends upon the same factors -- topography, water supply, and need. Nursery

ponds usually are smaller than rearing ponds because the fry are very small. The size of nursery ponds depends on the fish species being cultured. In fact, eggs and fry can even be kept in washtubs, oil drums or any other such container which holds enough water for the number of fry and is supplied with enough oxygen.

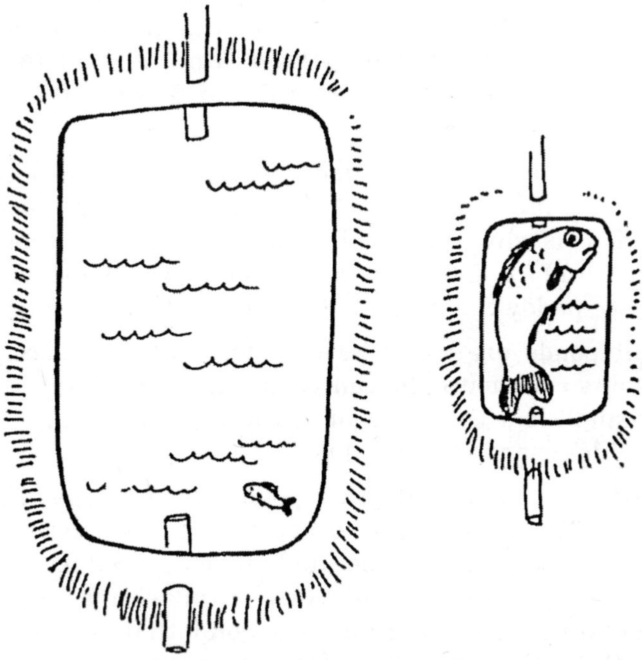

As the fish grow, they need more space. So rearing ponds are usually bigger than nursery ponds, and brood ponds are bigger than rearing ponds.

Sometimes a farmer will have to choose between one large pond or several smaller ponds. His site would allow him to decide either way.

Here are some advantages of small and large ponds:

Small Ponds:
- harvest easily and quickly
- drain and refill quickly
- treat for disease easily
- are not eroded by wind easily

Large Ponds:
- cost less to build per hectare of water
- take up less space per hectare of water
- have more oxygen in the water
- can be rotated with rice or other crops

For most farmers, a few small ponds are better than one or two large ponds. Farmers must also manage their agricultural crops, and it is difficult for them to manage large ponds. Also, most farmers Just do not have a lot of land. A good size for a single fish pond is probably between 1 and 5 ares (100 and 500 m^2).

Farmers are going to be most interested in working the fish pond into an already going farm as simply and easily as possible. This is why culturing fish in rice paddies is popular in some areas. In fact, fish ponds can be set up in almost any area where a rice paddy can be located --even on steep hillsides.

Small ponds are easier to care for and construct. As a farmer gains experience, he can go on and build larger ponds. Starting small is a good idea until the farmer feels he knows what he is doing and is successful.

Depth of Ponds. The depth of ponds depends upon the fish being grown. Fish species like different kinds of food, and the depth of the ponds affects the kinds of food produced by the pond. A common carp, for instance, eats worms and other bottom organisms and must have a pond that is not deeper than 2m. But when the carp are fry, they eat only plankton, the tiny free-floating plants and animals suspended throughout the water. So nursery ponds for carp fry are often only O.5m deep. (As mentioned before, eggs and fry can be taken care of in almost any container which holds enough water and has enough oxygen).

Other fish feed at other levels in the ponds depending on their life stage and on their own food preferences. A very deep pond will not produce as much food because the sunlight cannot light the water below a certain depth and the plankton will not be able to make oxygen for the fish (see water quality). On the other hand, a very shallow pond might be turbid, covered by water plants easily, and become very hot. Most pond owners make sure that the water depth at the edges of the pond is at least 75cm to discourage water plants. It is best if the pond is about 75cm deep at the shallow end and up to 2m deep at the deepest end. This will give the best results with most pond fish.

THE ONE-POND OPERATION If the farmer's site can only have one pond, his decision is easy. It is hard to breed fish when only one pond is available. Usually a single pond is used only for rearing fish from fry or fingerlings to market size. This is the case in small, backyard fish ponds that ore used to supply fish for only one family. A good minimum size for such a pond is l5m2 in area and lm deep. A smaller pond would probably not be worth the effort to build and maintain.

A single pond is stocked with the fry or fingerlings. For example, a pond of the size mentioned above could be stocked with 60 fingerlings. These young fish are cared for until they reach adult size. Then the pond is harvested (the fish are taken out). The pond area can then be prepared for a new batch of fish and stocked again.

One pond can provide a good food source for the family. However, rearing fish means that somewhere there must be a source of fry or fingerlings for use in the pond. The farmer must check his area carefully, so that he is sure the young fish are available before he builds one pond.

The source can be a river where he collects the young fish, or a local fish farm which breeds fish to supply farmers who have small ponds, or a government hatchery where the farmer can buy the young fish. If the farmer decides that he wants to breed fish in his pond, it Is possible to breed some fish inside small nets placed in the pond. A single pond, though, is usually used just for rearing fry or fingerlings to a good size for food and market.

While one pond usually means that the farmer is wise to concentrate on raising one batch of fish from fry or fingerlings to market size, he still must decide what kind or kinds of fish he will raise in his pond. He can raise one kind of fish alone (monoculture), or he can raise several kinds together (polyculture).

MONOCULTURE Monoculture is the culture of only one species (kind) of fish in a pond. It can be tilapia of one species, common carp, or any other single fish species.

Monoculture has some advantages. One advantage is in intensive fish culture practices, where fish are fed a lot of supplementary foods for fast growth. It is easier to give these foods If there is only one type of fish in the pond. Another possible advantage is that monoculture gives greater control over the age and sex of the fish. In monocultures, fish can be of all different ages and life stages, or they can be separated into fry, fingerlings or brood stock.

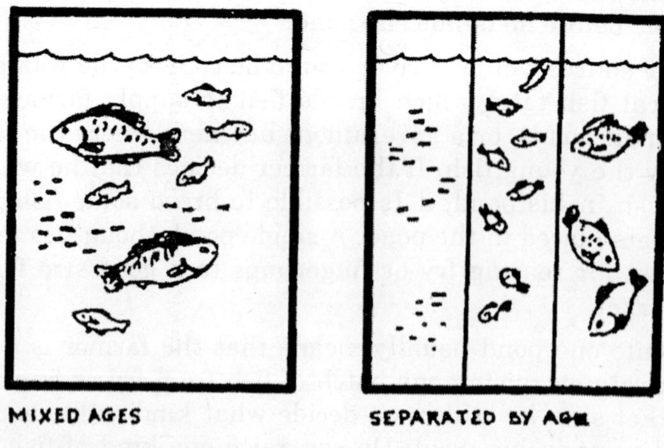

A monoculture allows a farmer who is unfamiliar with fish farming to get to know his one type of fish very well. And there is some advantage to this.

One disadvantage of a monoculture pond is that it is more likely for a single disease or parasite to kill all fish in the pond. Different fish are susceptible to different diseases. If only one fish type is present in the pond, a bad fish disease could easily infect and kill all the fish if it were not stopped in time.

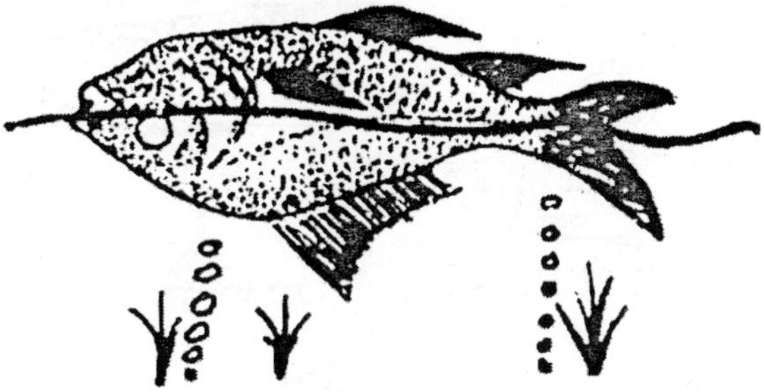

In monoculture ponds, fish are harvested selectively by using nets which have meshes of different sizes. For example, if the farmer wishes to harvest larger fish for market or breeding, the net will not catch or hurt the fry or fingerlings, because they are too small to be caught by a large-mesh gill net. This allows the farmer to keep his pond in operation and producing fish for food all year.

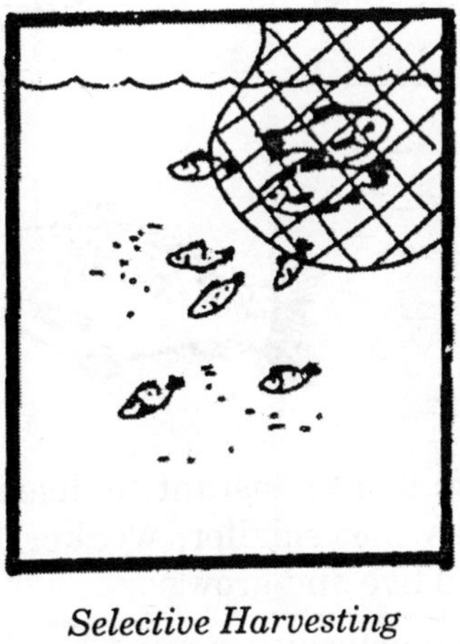

Selective Harvesting

Monoculture is the most common kind of pond culture. For a small fish farmer who is most interested in having a nearby, year-round supply of protein (and who does not have a lot of time or interest to give to the pond), a monoculture may be a very good idea.

POLYCULTURE

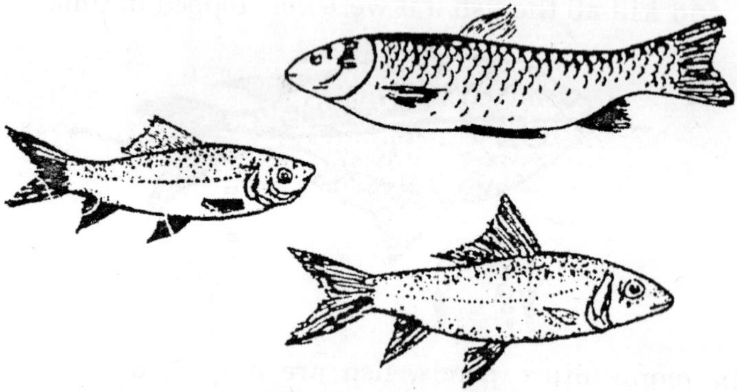

Polyculture is the culture of two or more fish species together in a pond. A good polyculture uses the natural food sources in a pond better: if the polyculture is mixed correctly, each of the species eats a different food from the pond.

Polycultures are more resistant to disease. Disease, if present, usually attacks the: smaller, weaker fish, and the healthier fish continue to live and grow.

Fish stocked in a polyculture must be able to live together. And living together successfully means that the fish put into the pond together do not all need to eat the same food. A polyculture can have fish of any size or age -- as long as a balanced relationship is maintained.

Some examples of polycultures are:

- fingerlings of two or more species stocked together in a fertilized pond and left to grow. A good mixture in this kind of polyculture is a mixture of Chinese carp -- silver, grass, and bighead carp stocked together. The silver carp eats phytoplankton; the grass carp eats pond vegetation; the bighead carp eats zooplankton.

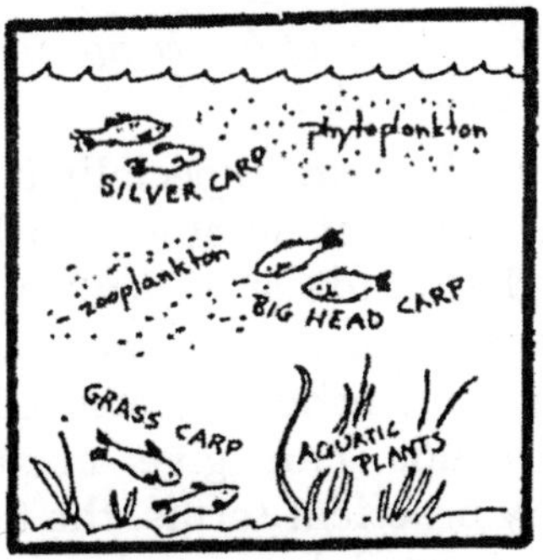

MIXED CARP FINGERLINGS

TILAPIA FINGERLINGS + CATFISH

- A few large fish (brood size) are stocked with fingerlings of another species in a pond and left alone. A good example of this is stocking tilapia fingerlings together with a few adult-sized *Clarias* catfish. The catfish feed on bottom organisms and serve as a population control on the fry that are produced in the tilapia ponds. Since one of the problems which can be associated with culturing tilapia is over population, this is a very complementary relationship.

- Another example of this type of polyculture is a stocking of any kind of fingerlings mixed with a few large grass carp for weed control.

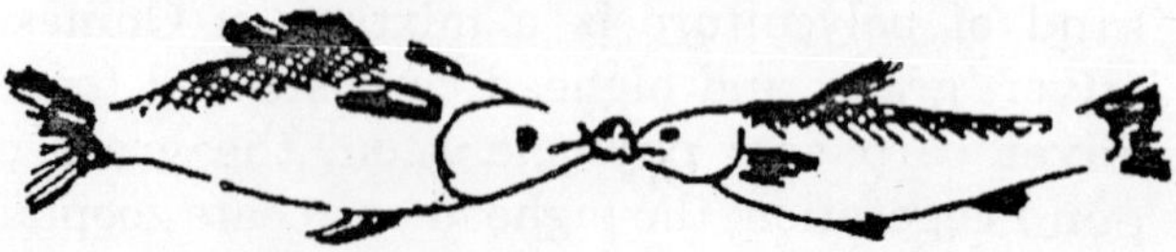

A farmer must be careful to avoid the problem of fish competing for food when he plans a polyculture

Polyculture is a good way to use a pond, especially if there is only one pond to use. A careful examination of local fish and their habits should tell a farmer what kinds of polycultures are possible in his pond. The important thing to remember is that the fish must not compete with each other. If stocked and managed correctly, polyculture ponds can give maximum production to a fish farmer. In very practical terms, the farmer could raise as much as three times more fish in a polyculture of three species than he can raise in a monoculture pond of the same size.

MONOSEX CULTURE A word should be said about mono sex culture, even though few farmers will choose or be able to choose this way of operation. Monosex culture means growling only one sex of one species of fish in a pond. When only males or only females are stocked in a pond, all energy of a fish goes into growth and not into reproduction.

An all-male stocking has faster growth rates than a mixed stock of males and females. So some farmers try to stock only males or females in a pond. One fish species that often is used in monosex culture is tilapia. Tilapia reproduce at a very small size, but when separated by sex, they do not develop their reproductive organs, yet continue to grow.

One way to stock a monosex pond is to separate the fish one by one according to sex during the breeding season. Often, at this time, fish change color, and it is easier to sort fish by sex. Then the fish can be grown to a larger size.

In another method, people have been trying to obtain fish of all one sex by putting two different species of tilapia into a pond. When these fish breed, they produce either a monosex culture or a sterile hybrid. Three crosses do now produce 100% male offspring.

Crosses of Tilapia which Produce 100% Male Offspring:

MALE	CROSSED WITH	FEMALE
Tilapia macrochir	X	*Tilapia nilotica*
Tilapia mossambica	X	*Tilapia nilotica*
Tilapia hororum	X	*Tilapia mossambica*

There are no crosses that produce 100% female offspring as yet. Males are preferred because they continue to grow during the breeding season, when there are no females present -- even though they (the males) continue to build their nests in preparation for mating.

Monosex culture is a valuable method of pond culture, but is usually difficult to do: the hybrid crosses are very new; hand-sorting fish by sex causes many of the fish to die from stress. Even if the fish are sorted without stressing them, one fish of the opposite sex that accidentally finds its way into the pond can ruin the whole monosex culture. So monosex culture is generally not practiced by small-scale fish farmers.

THE MORE-THAN-ONE-POND OPERATION A farmer who has a larger area to work with might wish to consider having two or three small ponds. Perhaps two ponds would be diversion ponds, and the third, a barrage pond fed by a spring. Perhaps the farmer has room for only two barrage ponds. He does not want to keep eggs and fry in the ponds because it is harder to protect eggs and fry in barrage ponds. This does not mean he cannot breed fish. He can keep eggs and fry in an oil drum, washtub, or anything else as long as the water is clean and contains plenty of oxygen.

With three ponds, one pond can be the rearing pond in which fingerlings are raised to market size; one can be used to keep brood stock; and the third, and perhaps the smallest, can be used as a nursery pond where the eggs hatch and the fry grow to fingerling size. If the farmer does not plan to breed fish, then he can use all three ponds as rearing ponds. He should not do this, however, without thinking ahead to the harvest and making plans for marketing the fish he will grow, or preserving the fish for sale or use later.

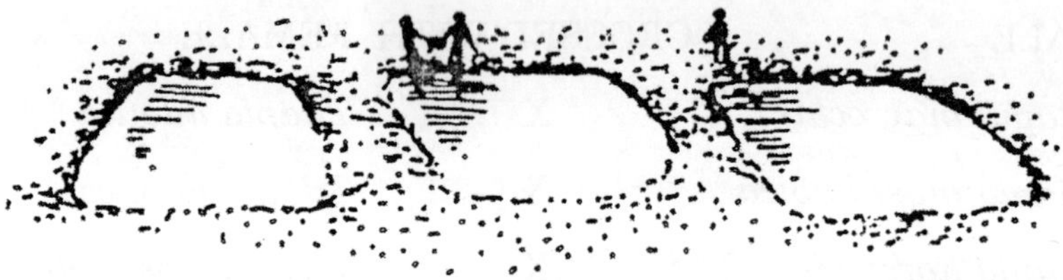

The major difference between a large farm operation and a small one my be only the number of ponds. Three ponds is enough to have a full-fledged operating fish farm which includes breeding, selling fry and fingerlings to other farmers, and raising fry and fingerlings to market and brood size. Once the farmer is a skilled pond manager, these ponds should do well and provide a good return on his investment.

Until the farmer is experienced, however, it is better for him to start with small efforts and a smaller operation. Small pond failure is not as severe. Once the ponds are working well, the farmer can expand and build more and/or larger ponds. But he should be encouraged to start small. There are a lot of factors in fish pond management that are learned best by experience. But a bad experience will discourage, rather than encourage, the pond owner.

A FINAL WORD ON PLANNING PONDS Good planning is a must for a successful fish pond operation. It is during the planning process, before any money or a lot of time and energy is spent, that many problems can be solved.

The farmer should keep in mind while planning that ponds do not have to have expensive equipment in order to work well. Far more important than the equipment are 1) an understanding of the general principles involved, 2) the selection of a fish or fishes that will do well in his pond (see next section, "Selection of Fish"), and 3) good daily management of the pond (see section 6,"Managing the Pond").

3 Planning: Selection of Fish

The farmer now has a firm idea of his site and the types of ponds it is possible for him to build. He also should know what he wants to do with his ponds -- raise fish for food or run a fish-marketing business. Now he must consider very carefully what type or types of fish he is going to raise in his ponds. The success of the pond depends upon choosing the fish that will grow best in the type of ponds and conditions that a farmer is planning.

The following pages give some: 1) general information on characteristics of fish, and 2) detail about certain fish which have proved to be good pond fish and why. This information should serve as a guide to -a farmer trying to decide which fish will do best in his ponds.

Characteristics of Fish

The major body parts of all fish perform the same functions, and they are located in about the same places on any different fish's body. But the size, shape, and color are often different, and these differences help tell the fish apart. Knowing how a healthy fish looks is important.

All fish have a tail consisting of the *caudal peduncle* and the *caudal fin.* The fish's fins help it steer through the water and hold it upright in the water. Often a sick fish cannot steer or flops over on its side. Other fins on the body include:

- *Pectoral* --usually located on the sides of the fish behind the head.

- *Pelvic* --usually located towards the rear of the body where the hips would be if the fish were a four-legged animal.
- •*Dorsal* --runs along the top of the fish. May be single or double. The second dorsal fin is sometimes called the *soft dorsal.*

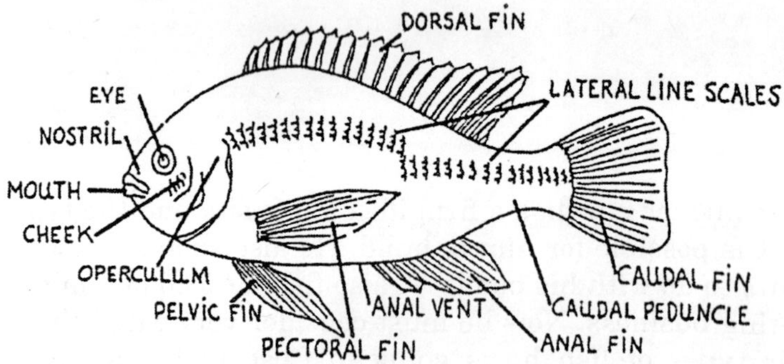

Parts of the tilapia, one of the best of tested pond fish.

- *Anal* --usually located right behind the *anal vent* (anus) on the rear bottom end of the fish.

Most fish have *eyes,* but even with eyes fish cannot see very well. All fish have *gills.* The gills are covered by a flap called the *operculum.* The gills are extremely important. Fish take in water through their mouths. The water is then passed through the gills which remove the oxygen and nutrients from the water. The water is then passed outside of the body of the fish through the gill slits.

It is possible to tell a lot about a fish's health and eating habits by looking at its gills. Fish with many, many feathery gill rakers and few if any teeth eat the smaller foods in the pond. Fish with few and larger gill filaments eat the larger particles from the pond. Healthy gills are a bright red color. If the farmer sees fish with gills that do not have this healthy red color, or have white spots all over, for example, he will know that fish is not healthy and should not be bought or placed in his pond. Or if the fish is already in his pond, he knows he must take steps to get rid of the disease before it troubles more fish.

Other identifying parts that all fish have are the *mouth,* the *genital openings* (to reproductive organs), and the *lateral line.* The lateral line is a small line of nerve cells which runs along the length of the body about midway on the side of the body. Sometimes the lateral line is covered by a layer of scales; sometimes it is a different color than the rest of the body. In any case, the lateral line is an area of sensitivity that helps the fish feel pressure and temperature changes in the water around it.

Some fish, like catfish, also have *barbels,* small projections that hang down from the sides of the mouth. Barbels help the catfish sense its surroundings, find food and attract small fish to the catfish so that it can eat them.

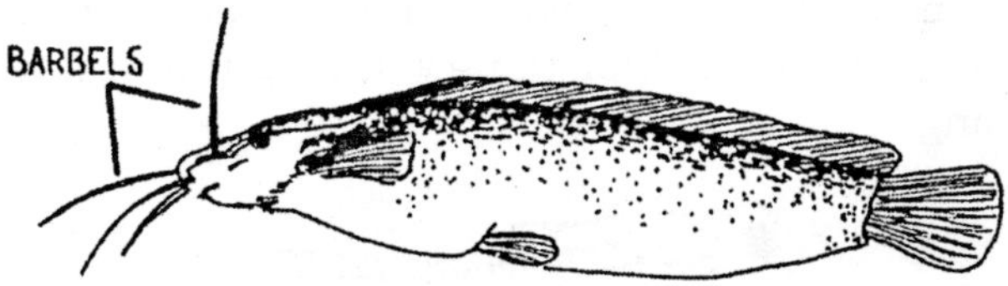

When a farmer breeds fish he will want to be able to tell the difference between male and female fish. This can be difficult with some fish. However, some fish change color in the breeding season (tilapia, for example), so they are easy to identify by sex. Some fish can be classified according to the color and size of their genitals. The separation of fish by sex is best learned by actual experience in the pond.

Sex organs of tilapia

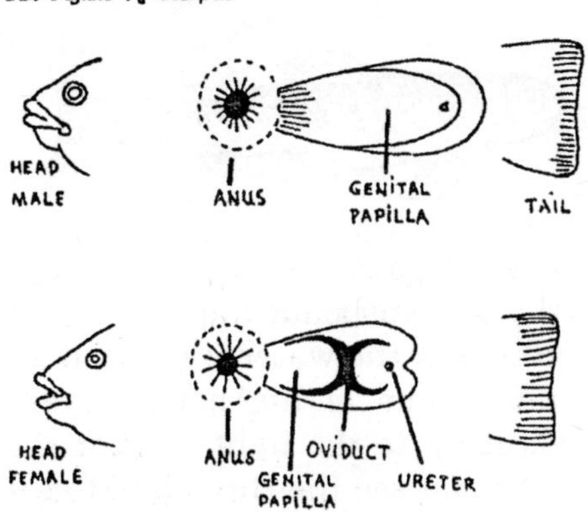

When the farmer goes to buy fish, he must already know what healthy fish look like. It is very important that he be as familiar as possible with each of the fish he decides to raise. He must know the characteristics of that fish and its life cycle, Its eating and breeding habits, etc. The farmer who begins any fish pond enterprise without having this kind of information is inviting failure. And if it is a new venture, it is particularly important that the farmer's first effort be as successful as possible.

The Life Cycle of Fish

Fish start life as fertilized eggs. The eggs grow and then hatch into small fish, called fry. The fry are attached to the yolk sac which is the leftover part of the egg they hatched from. The yolk sac provides food for the fry during the first few days after hatching.

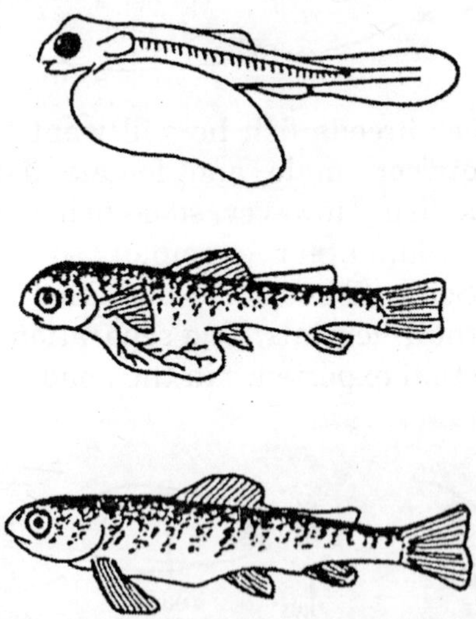

After the yolk sac is gone, the fry searches for food in the water. All fry eat the tiny suspended and swimming plants and animals called plankton in the water. Plankton are hard to see, but if a farmer puts some of his pond water into a glass container and holds it up to the light so that the light shines through the water, he can see the tiny plankton floating in the water. The length of the fry stage depends upon the species of fish. Usually a fish is a fry at least until the yolk sac is absorbed. Fry range from 2mm to 30mm in length. This growth process can take 2 to 6 or 8 days depending upon the type of fish.

As the fry grow bigger, they are called fingerlings. They are called fingerlings because at this stage of the growth cycle, they are about the size of a person's finger. Fingerlings vary in size -- from 4-10cm. Above 10cm, the fish is better called a post-fingerling. The adult fish ranges in size; some can be as large as 2m long and weigh 22 kg. An adult fish is a fish which is sexually mature.

Fingerlings have different eating habits from fry; they are now much bigger and can eat larger pieces of food. As fingerlings, the fish begin to show that they like certain foods better than other foods. Each kind of fish chooses its own kind of food, depending upon his needs and what is available. For example, a carp fry will eat plankton; as a fingerling, the carp eats pieces of decayed matter and insect larvae; as an adult the carp will eat plankton, decayed matter, insect larvae, worms, snails, and almost anything that is on the bottom of the pond. Common carp, for example, are called "bottom feeders," because they eat food from the bottom of the pond.

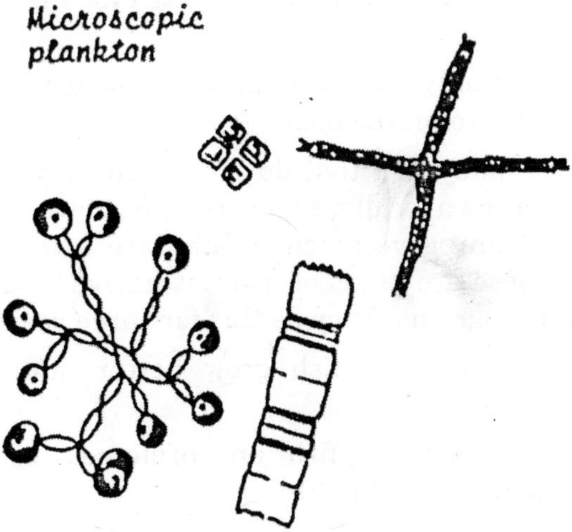

The food preference does not always change as the fish grows. Some fish, like the silver carp, eat plankton their whole lives. When the fish reach adult size, they will sexually mature in the right conditions. Brood fish are sexually mature fish which are chosen as good fish to breed (spawn), produce eggs and begin the whole cycle again. This is called the life cycle of a fish.

Knowing how the fish in the pond grow, and the foods they require at each stage in the life cycle is very important for good pond management.

Choosing Pond Fish

Choosing fish to grow in ponds can be difficult. A good pond fish has certain characteristics which help it grow successfully in ponds. There are some fish which will not adapt to pond conditions and cannot be used in pond culture. A pond is very different from a natural waterway:

- There is usually no water flowing through a pond. Some fish need to live where there is quite a bit of current in the water, rather than in a quiet pool of water.
- The food that is already in the pond is all that is available to the fish, unless extra food is put in by the farmer.
- There is only a certain amount of water and pond area in which to move about.

There are many fish that do grow well in ponds. Some of these are fish, grown locally; some are fish grown in other parts of the world. Many governments today are introducing exotic fish species (these are kinds of fish not native to that country) into fish pond programs. They do this for three reasons:

- Some introduced fish grow better and faster than native fish.
- Some introduced fish are preferred by people for eating (over local fish).

- The offspring of a cross between a local fish and an introduced fish sometimes grow faster and taste better than either of the parent fish (this is called *hybrid vigor).*

But exotic fish must be watched and used very carefully. They must not escape into local waters. Some exotic fish which escape create problems in natural waters when they begin to compete with local fishes for food. Also, introduced fish can carry diseases or parasites that are fatal to native fishes.

There are certainly a number of fish in the natural waterways of your area which will grow well in ponds. Native (local) fish are usually easier to use because they are adjusted to local water and climate conditions.

If at all possible, farmers should be encouraged to start their ponds using a tested pond fish which is locally available and is well-liked by people in the area. It can be a fish from the list given here or one chosen from a list prepared in your area. The important points are that the farmer be able to sell any fish he wishes to sell, that the fish can grow in ponds, and that there is brood stock available locally.

Fish Used in Pond Culture

Here are some characteristics that good fish for pond culture will have. Certainly it my not be possible for a farmer to determine whether a certain fish has all these characteristics right away, particularly for those local fish not discussed in detail here or those newly Introduced to pond culture. But good pond fish all have certain characteristics: the more certain a farmer can be that the fish he chooses to raise fit these descriptions, the more sure he can be of his success. Good pond fish are:

- available locally
- able to reproduce (breed) naturally in your area
- able to live in a confined space (the pond).
- able to find the right foods in ponds.
- fast-growing.

- relatively free of parasites and diseases.
- known and liked as a food fish in the area.

Some fish that fit these criteria for good pond fish and are now grown in ponds all over the world are named here. Though they all are grown in ponds, each has certain characteristics which mean that it will grow better in some kinds of ponds better than other ponds. Of course, these fish are not the only fish that can be used in ponds, But they are named here because they have bean tested in ponds, and they can grow well under pond conditions. All of these fish are warm water fish.

SCIENTIFIC AND COMMON NAMES OF FISH USED IN POND CULTURE

Please note: Each fish has a scientific name which is always the same. The common name, however, can be different from one country to the next. It is a good idea for anyone who works with fish to know the scientific name.

	Genus -species	**Common name**
1	*Anguilla japonica*	eel
2	*Aristichthys nobilis*	bighead carp
3	*Barbus gonionotus*	tawes
4	*Carassius auratus*	goldfish
5.	*Carassius carassius*	crucian carp
6	*Catla catla*	catla
7	*Chanos chanos*	milkfish
8	*Cirrhium molitorella*	mud carp
9	*Cirrhina mrigala*	mrigal
10	*Clarias batrachus*	catfish
11	*Claria macrocephalus*	catfish
12	*Ctenopharyngodon idellus*	grass carp
13	*Cyprimus carpio*	common carp
14	*Helostoma temminchi*	kissing gourami
15	*Heterotis niloticus*	–

16	*Hypophthalmichthys molitrix*	silver carp
17	*Labeo rohita*	rohu
18	*Mugil cephalus*	mullet
19	*Mylopharyngodon piceus*	black carp
20	*Osphronemus goramy*	gourami
21	*Serranochromis robustus*	–
22	*Tilapia macrochir*	tilapla
23	*Tilapia melanopleura*	tilapia
24	*Tilapia mossambica*	tilapia
25	*Tilapia nilotica*	tilapia
26	*Trichogaster pectoralis*	snakeskin gourami
27	*Trichogaster trichopterus*	three-spot gourami

Following is specific information on some of the more popular pond fish.

COMMON CARP

The common carp, *Cyperimus carpio* is a favorite warm water pond fish. Common carp are used as a pond fish because they:

- spawn easily in ponds.
- do not get sick easily.
- tolerate wide ranges of temperature and pH (factors of water quality discussed in detail later).
- eat all kinds of food, from zooplankton to decaying plants.
- have a very good growth rate.
- accept supplementary foods.

Common carp generally are a grey-green color. However, they also can be gold, yellow, orange, pink, blue, green, or grey. They spawn all year round in warm waters, and they can be made to spawn by the pond owner if they do not spawn naturally. Common carp are good to eat when they are cooked properly. They can be grown in ponds by themselves (monoculture) or in ponds with Chinese or Indian carp (polyculture).

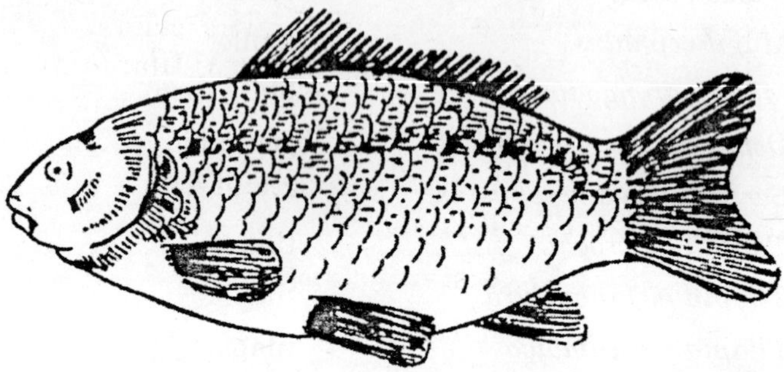

Some of the yields gotten in various countries by stocking common carp in monocultures are shown in the following table.

Country	Culture methods	Yield, kg/hectare
Czechoslovakia	Growth in ponds with ducks	500
Guatemala	Intensive culture in ponds	4,000
India	Natural growth in ponds	400
	Growth in ponds with management	1,500
Indonesia	Intensive culture in ponds	1,500
Japan	Intensive culture in ponds	5,000
Nigeria	Commercial culture with fertilization and feeding	371-1,834
Philippines	Intensive culture in stagnant water	5,500
United States	Intensive pond culture with inorganic fertilization	314

Source: Bardach, *et al.* (1972)

Conclusion: Common carp are a very easy fish to breed, keep, and harvest, so a fish pond that relies on common 4arp will probably do well. Common carp are a good fish for a farmer to use for his first effort. With good management, common carp will continue to produce healthy eggs and fry until they are too old (above 5 years of age).

TILAPIA

The *Tilapia* genus (family Cichlidae) contains at least 14 species, which are all good pond fish. The color of the fish differs only slightly depending upon species; tilapia are generally dark brown to black in color. The most common species grown in ponds is the *Tilapia mossambica,* also called the Java tilapia. It has been introduced throughout the world and is easy to find in most places. Tilapia:

- are hardy fish, resistant to disease.
- breed easily in ponds.
- grow rapidly.
- taste good.
- can withstand wide temperature ranges.

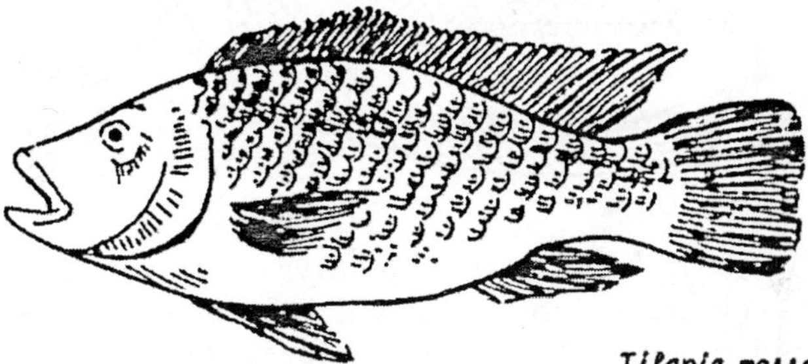

Tilapia mossambica

Tilapla are herbivorous: some species eat higher plants; some eat phytoplankton. Both the Java tilapia and the Nile tilapit *(Tilapia nilotica)* do well in very enriched waters (waters polluted by sewage). All tilapia have slightly different eating habits, depending on the species.

Tilapia reproduce every month or so, once they become sexually mature. They then take very good care of their own eggs and fry in ponds. If the farmer plans to breed and raise fry, this fish is a good choice because the fish themselves take care of the fry at a stage where many fish of other species die easily. The major problem with raising tilapia in fish ponds is that they become sexually mature at a small size, and begin to reproduce instead of to grow further. It may be necessary to separate the tilapia by sex before they are old enough to reproduce. Or it may be necessary to introduce catfish into the pond to control the population of small fish.

Conclusion: Tilapia species have many possibilities for pond culture. Their fast growth rate, ease of breeding, good taste and hardy bodies make them a good choice, particularly for the first-time fish farmer.

CHINESE CARPS

Other kinds of carp, besides the common carp, often are grown in ponds. Most commonly used are the Chinese carps. Some of these are:

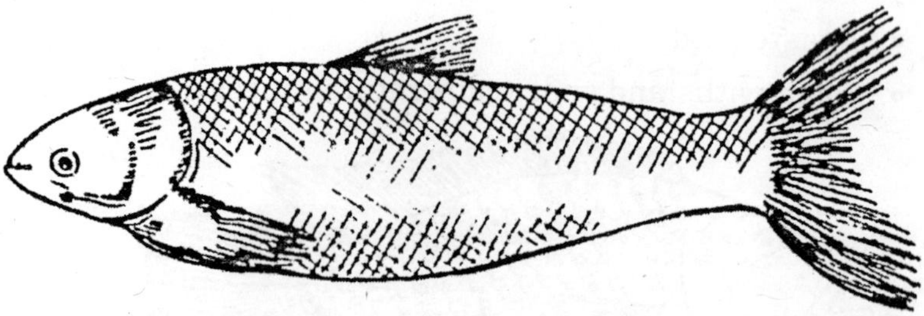

- Silver carp (*Hypophthalmichthys molitrix*). This fish eats phytoplankton, but will accept rice bran and bread crumbs. The silver carp gets its name from its silver color. It has very small scales.

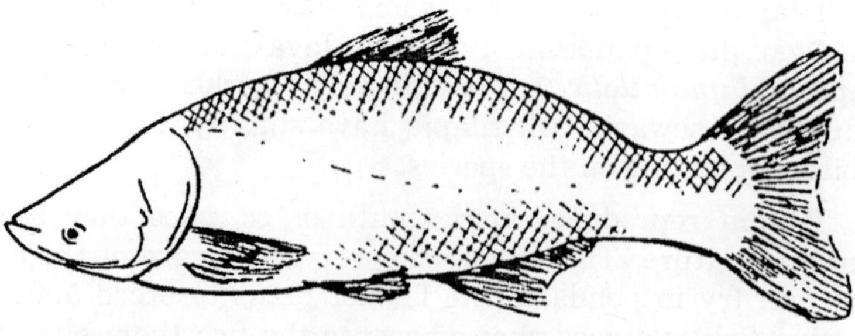

- Bighead carp (*Aristichthys nobilis*). This fish feeds mainly on zooplankton. It is a dusty green color on top which fades to a pale green color on the abdomen. It also has small scales.

- Grass carp *(Ctenopharyngodon idellus).* This fish is an herbivore and eats water vegetation (but also will eat almost anything). The grass carp is also silver-colored, but has a darker grey area running along the top of the body. It grows larger in size and has larger scales than a silver carp

Other chinese carps like the black carp *(Mylopharyngodon piceus)* and the mud carp *(Cirrhina molitorella)* are bottom feeders. This difference in eating habits is very important in fish pond culture. It is the reason why polyculture, or growing a number of fish species in one pond can be successful. When one kind of fish is stocked alone in a monoculture, the foods in the water not eaten by that type of fish are wasted. In a polyculture of the above three species of Chinese carp, for example, three kinds of food are being eaten.

The following table gives some examples of polyculture mixes and of how many fish of each kind can be stocked in a pond. For example, Pond I is stocked with silver, bighead, grass and common carp.

STOCKING RATES OF CHINESE CARPS IN PONDS 3 TO 7 METERS DEEP IN KIANGSU PROVINCE, CHINA

Species	Weight of Yearlings, grams	I	II	III	IV
Silver and bighead carp	500	4500	4500	9000	9000
Grass carp	500	600	—	3000	—

Black carp	500	—	450	—	3000
Common carp	200	200	200	200	200
Total		5300	5150	12200	12200

Source: Bardach, et al. (1972)

The preceding table shows polyculture mixes: as you can see, common carp can also be used in polyculture with Chinese carp. Chinese carp are grown in ponds because they grow well in polycultures, and they are very good to eat. The silver carp grows faster and is tastier (according to some farmers) than common carp. The grass carp is most often used to control weeds in the pond. In fact the grass carp does a better job of weed control than do chemicals. The grass carp is perhaps the most interesting of the Chinese carp and Is now being studied by scientists in many countries to find better ways of breeding it in ponds.

A farmer might run into problems raising Chinese carp -- if he does not look into his local situation very well. Farmers will have to have a source of Chinese carp fry from a government hatchery or a local breeder before trying to raise Chinese carp. The carp only breed once a year, and then, inmost cases, only with help from man. Also, Chinese carp are very susceptible to diseases. Then, because they are delicate fish, they must be handled very carefully, or they will be injured.

Conclusion: A farmer just beginning a fish pond probably would not want to breed Chinese carp, but he certainly should be familiar with these fish and how they might help his ponds. For example, even two or three large grass carps placed in a pond with many fish of one other species could be valuable for keeping a pond balanced.

INDIAN CARP

There is one last group of carp often cultured in ponds. These are the Indian carp. Indian carp are further divided into minor and major carp. The major carp of India are the catla (*Catla catta*), the rohu *(Labeo rohita),* and the mrigal *(Cirrhina mrigala*). The minor carp are the reba, the bata, the sandkohl, and the nagendram fish. The Indian major carp will not spawn in standing water, so special ponds are built in India to provide a flow of water for these fish, who must have running water in which to spawn. The Indian carp can be made to spawn by man, but this is a difficult process (see "Manacling Brood Stock"). However, there seems to be no reason why the Indian carp cannot be spawned in ponds in places where ponds can be constructed to provide constantly running water.

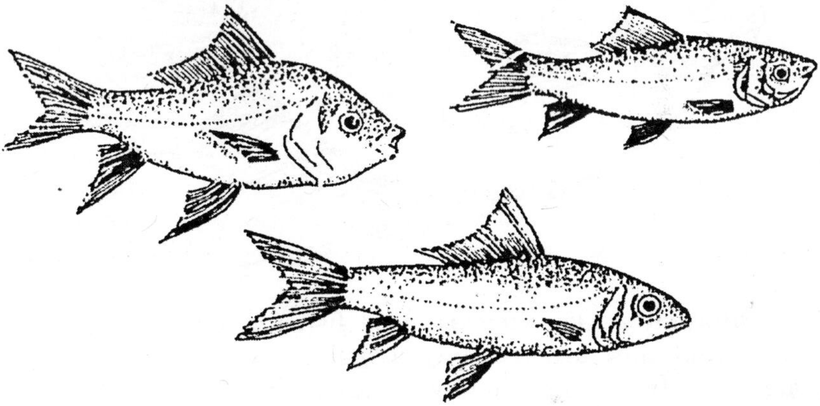

Conclusion: A farmer who has only a small pond should not try to breed Indian carp. Indian carp can be grown in polycultures with common carp, but are not as good or fast growing in ponds as the Chinese carp. Indian carp are also susceptible to many diseases. This is a fish for an experienced fish farmer who is interested in, and able to, experiment.

GOURAMI

The gourami *(Osphronemus goramy)* is a very good pond fish. It is originally from Indonesia, but now is grown all over Southeast Asia. Gourami possess an accessory air-breathing organ, which means that they can survive in waters that are low in dissolved oxygen. This makes it *an* important fish in areas where the temperature remains high and there is little water for certain periods of the year. Gourami spawn all year round in warm water conditions. Gourami:

- spawn easily all year round in warm waters.

- taste good.
- are easy to breed.
- accept a variety of foods.
- are hardy.

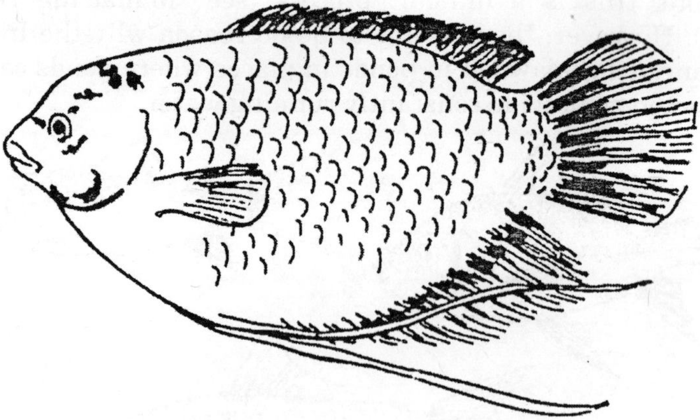

Conclusion: Gourami are good fish for a first-time fish farmer. And they are certainly a fish to be considered very thoughtfully by farmers who live in areas that remain very hot and dry for periods of the year. The gourami is used to these conditions, and there are other pond fish which would not do well at all under these conditions.

CLARIAS CATFISH

Clarias catfish are found throughout Asia, India, and Africa, as well as the Middle East. The species most often used as pond fish are *Clarias macrocephalus* and *Clarias batrachus*. *Clarias macrocephalus* is preferred for its good taste; *Clarias batrachus* grows faster.

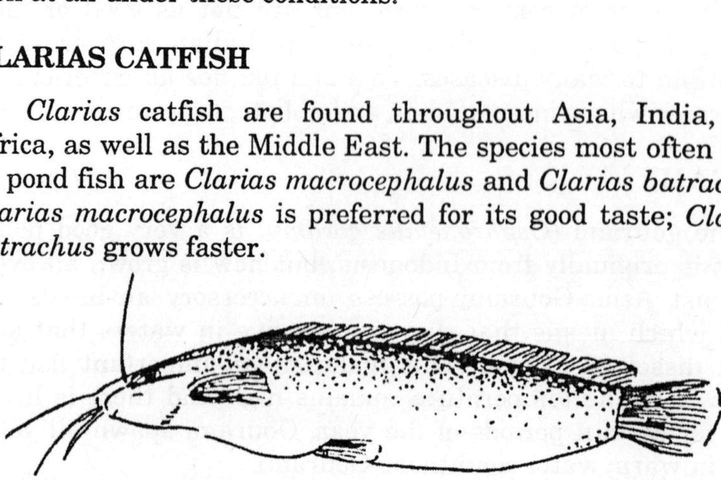

These catfish have accessory air-breathing organs; they can even crawl out of ponds to look for food. Because they can live in shallow ponds, these catfish are sometimes used in culture with rice (see paddy culture). They are scavengers, which means they will eat just about anything. However, they prefer to eat worms, snails, and other fish. They are often used in polycultures with tilapia where they serve as predators on the very small tilapia. They will eat supplementary foods, and give very high production in ponds. In Thailand, *Clarias* catfish yield about 97,000kg/ha when they are fed supplementary foods. These catfish are hardy: they sometimes get external parasites, but these do not kill the fish.

Conclusion: The catfish are another good fish to be raised in areas where high heat and long dry spells are found. They are good to eat. easy to keep, and can be used in ponds in a number of ways. Certainly a farmer who already cultures paddy rice might be interested in considering adapting his paddy to catfish culture.

TAWES

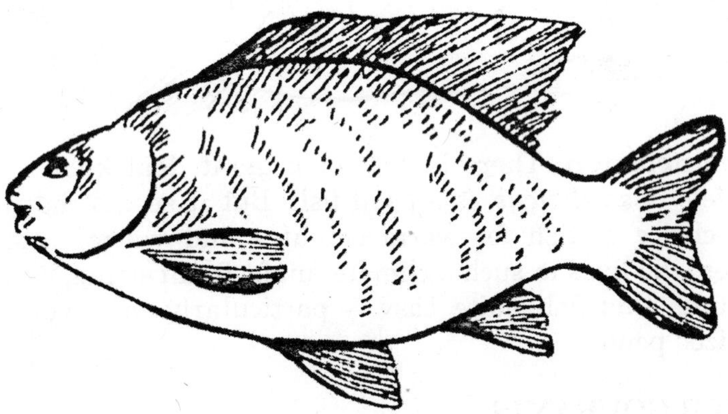

The common name tawes is applied to three species of fish -- *Barbus gonionotus, Puntius javanicus,* and *Puntius gonionotus.* These fish usually are used in fish ponds for vegetation control, in poly cultures with Chinese carp. Tawes are able to spawn all year round, but they most often spawn in the rainy season. Tawes need well-oxygenated water with a strong current to spawn. Tawes feed on soft water plants, but will also take rice bran. There is not a great deal known about the tawes at present, but It can be used in polycultures when the grass carp is not available.

Conclusion: A farmer starting a polyculture certainly might be interested in using this fish. However, first-time fish farmers with limited space would not want to try breeding this fish.

HETEROTIS NILOTICUS

The *Heterotis niloticus* spawn easily in ponds. The mature fish will build a grass-walled nest in the weeds at a pond's edge and spawn inside this nest. They spawn when water is low and very warm, at the end of the dry season. The mature fish feed only on plankton, but in a pond they will accept supplementary food. This fish has a swim bladder which can serve as an accessory air breathing organ.

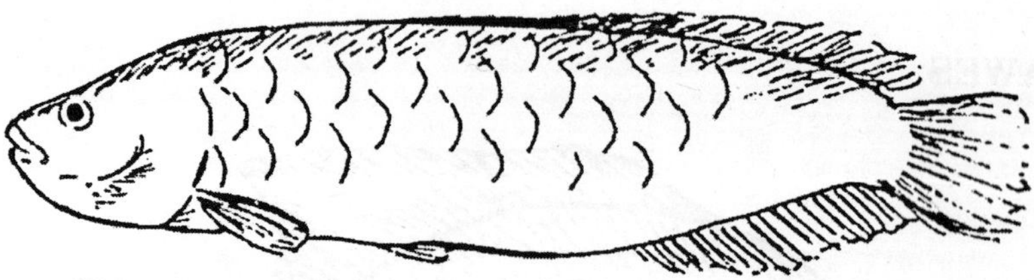

Conclusion: There is not yet a great deal known about the *Heterotis niloticus* as a pond fish. But it seems that it is a good choice of fish for warm climates and warm waters. A farmer who lives in such a climate might find raising, and even breeding, this fish quite easy -- particularly in a very well-fertilized pond.

OTHER GOURAMIS

These are the snakeskin gourami (or Sepat Siam -- *Trichogaster pectoralis),* the three-spot gourami *(Trichogaster trichopterus),* and the kissing gourami *(Heleostoma temminchi*). All of these good. And they breed easily in well-oxygenated, warm water. They do require a pond which has a good growth of vegetation (particularly *Hydrilla verticillata*).

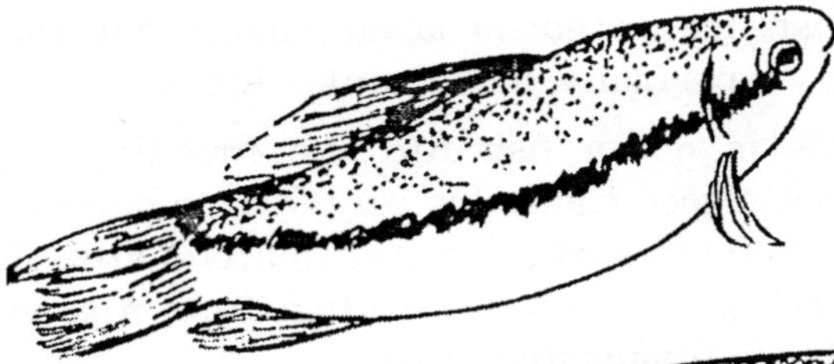

Snakeskin gourami

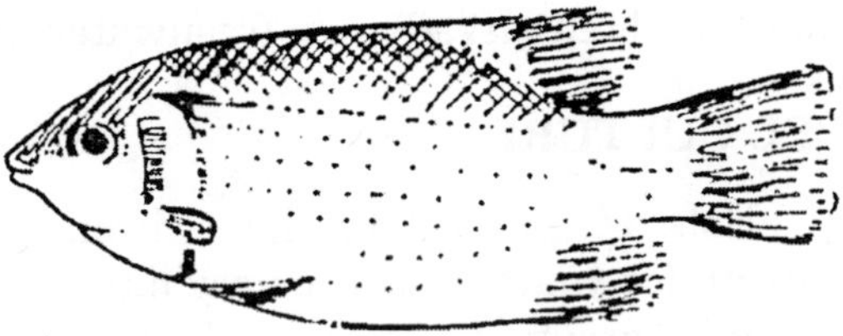

Kissing gourami

Conclusion: In a pond situation such as that outlined above, these gouramis are easy to breed and raise. They are a good fish to use in polycultures with other gouramis, tilapia, and common carp.

MILKFISH CULTURE

The milkfish *(Chanos chanos)* can be raised in freshwater even though it is primarily a brackish water fish, and will not breed in ponds. The fry are caught along the shoreline at breeding season (the rainy season) and transferred to freshwater ponds. Milkfish culture is done for the most part in the Philippines and in some other Southeast Asian countries, like Indonesia and Taiwan.

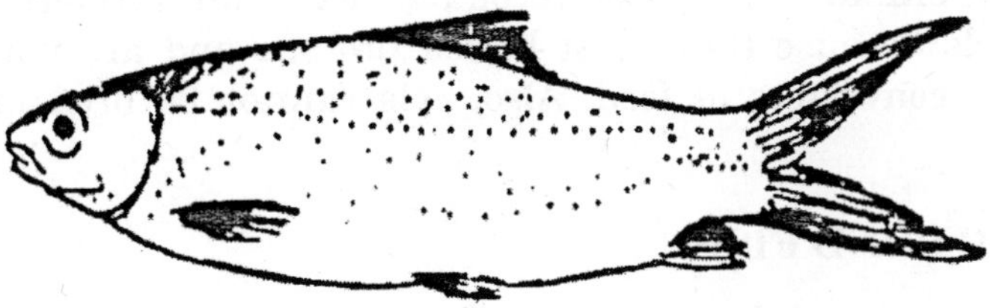

Adjusting (acclimatizing) the fry from the salt water to the freshwater pond is hard to do; many fish die if the adjusting process is not done well. Therefore, milkfish usually are cultured in brackish water ponds only; the use of milkfish in freshwater ponds is not widespread. Milkfish feed on a complex of bottom algae, and, recently, it Is reported they also feed on phytoplankton. Milkfish are prized for their beauty and their good taste, though they have many, many small bones.

Conclusion: This is not a fish for the first-time fish farmer. In fact, it is not a good choice for any farmer unless he has a saltwater pond; is interested in trying to acclimatize the fish to a freshwater pond; or can buy milkfish from a source that has them already in a freshwater pond.

EEL CULTURE

Eels (*Anguilla* sp.) have been cultured in Japan and Taiwan for years. Eels are very much a luxury food and are not normally grown alone in ponds outside of these two countries. The eels are grown in ponds in polyculture with other fishes and are particularly useful in polyculture with species of tilapia because they eat the smaller tilapias. The eels used in Taiwan (*Anguilla japonica)* spawn in the sea and the fry (called elvers) swim upstream and are collected by dealers. Eels must be fed supplementary feeds like pellets made of trash fish.

European eel

Conclusion: It is not recommended that farmers work with eels because they must be fed protein and are not very efficient converters of food. Also, eels cannot be bred in fish ponds.

OTHER POND FISH

Some other fish grown in ponds are the goldfish (*Carassius auratus*), the crucian carp *(Carassius carassius)*, and *SeitzunocdWmi4 Aobuaaa.* Any of these fish can be grown in polycultures with Chinese, common carp, and tilapia.

Conclusion: The use of one of these fish in a pond stocked with other, more important fishes, results in an increase in yields of both species. In polycultures these species can utilize other food sources and also act as predators and weed controllers.

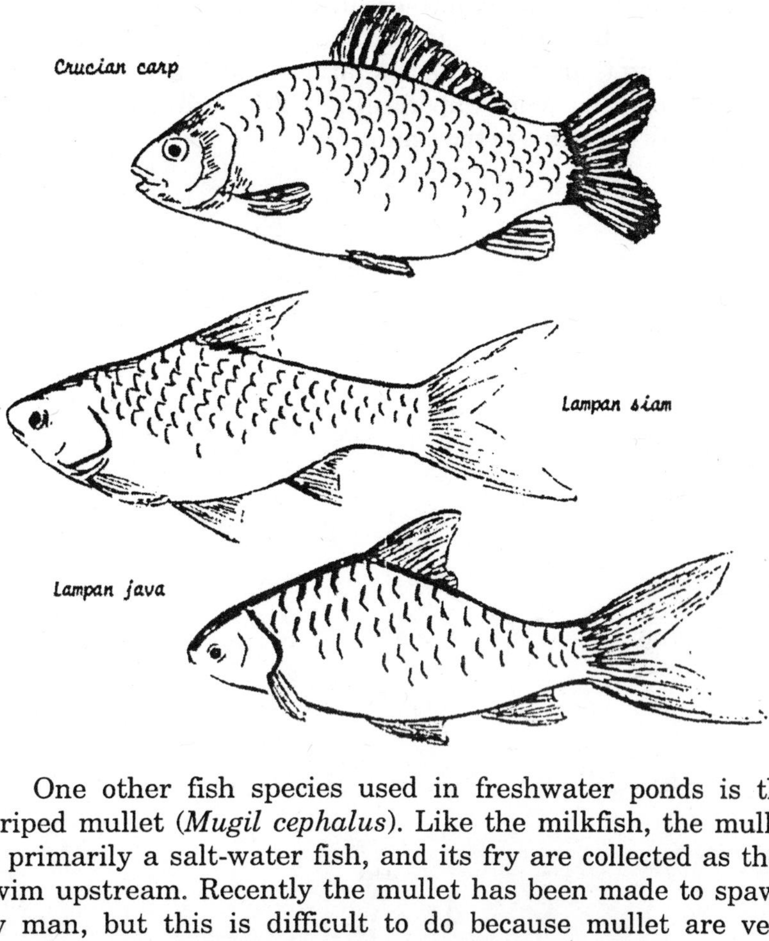

One other fish species used in freshwater ponds is the striped mullet (*Mugil cephalus*). Like the milkfish, the mullet is primarily a salt-water fish, and its fry are collected as they swim upstream. Recently the mullet has been made to spawn by man, but this is difficult to do because mullet are very sensitive to handling. However, mullet can survive in wide temperature ranges and are herbivores, so some farmers may want to try mullet.

A CLOSING NOTE ON FISH

All these fish have been and are now being cultured in fish ponds around the world. However, as stated before, they are not the only fish which can be grown in ponds. In every area there are a number of fish in natural waters that could be grown in fish ponds. So you might find it a good idea to experiment with local fish in your ponds, to find those fishes that might be available to farmers in your area for use in their ponds. It is better for an extension worker to do the experimenting than it is to have a farmer risk wasting his time or money, or even more importantly, risk failure. If a farmer fails, he may not want to try again.

4 Fish Pond Construction

Construction of a large pond can be very expensive if labor is hired, machines are used, and expensive equipment is rented. For example, in the Philippines, a one-hectare pond having two concrete gates and walls 3m high x 3m wide recently cost US$ 1,522.56. Another pond, about 100 m x 25 m, with only a Rivaldi valve cost about US$ 680.

An interesting fact about fish pond construction is that whether the pond is large or small, expensive or inexpensive, ponds are all very much the same. A larger, more expensive pond will not necessarily be a better pond.

Here is an example of a good beginning for a new and small fish farmer:

> A "backyard" fish pond was planned and sited very carefully by a farmer. The pond was dug by the farmer and constructed with bamboo pipes for water inlets and outlets. The construction itself cost no money. The farmer's only expense was a supply of fingerlings purchased from a nearby market. This fish pond, managed by the farmer and his family, produced enough fish for the family and some extra income from fish sold or bartered for goods needed by the farmer. The family ate well and suffered no major illnesses during the year.
>
> Next year, the farmer plans to add another pond and to produce more fish for market. He will add a Rivaldi valve or a wooden monk to this new pond, because either of these will make ongoing management somewhat easier, now that there will be two ponds to manage (Thc

bamboo pipe sometimes got clogged. This was no problem to correct when there was only one pond. But it would take up needed time in a two-pond operation). Whichever the farmer chooses, the valve or the monk, he will make it himself with materials found locally, using money from the sale of his fish.

This farmer began his operation well. He started small and worked into a larger operation. However, even for the larger fish farm, he planned an expansion which was within his means. This kind of careful planning increases the farmer's chances of success --and yours. And the scope of the project is something he can undertake on his own. He gained the knowledge and experience that he needs to expand his operation.

The following section presents a range of ideas for constructing fish ponds. The farmer can pick a combination of construction possibilities which best fit his own needs and resources.

IMPORTANT: Stress that the "right" way in any situation is the way which:

- the owner can afford
- the owner can manage easily
- fits the owner's needs most completely

Construction should begin only after careful planning such as that outlined in the preceding sections or "Planning."

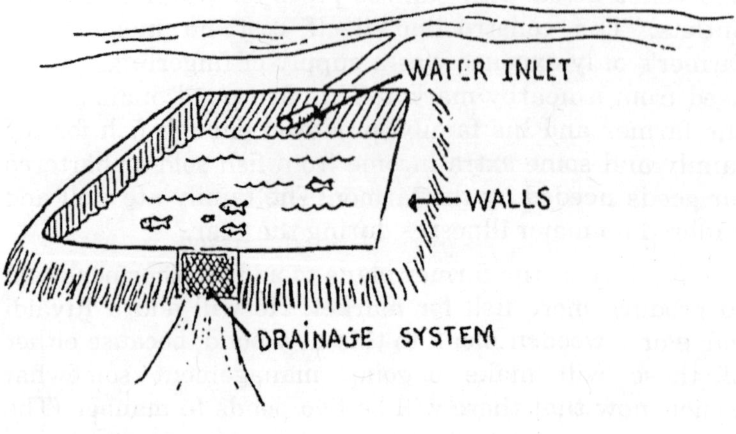

A fish pond has three main parts: the walls, the water inlet, and the drainage system. Walls are also called dams, dikes, levees, or bunds. This manual uses "walls." Whatever they are called, walls hold the water in the pond. They can be built using soil taken from inside the pond, or they can be built with soil taken from another place. They must be strong enough to withstand the pressure of all the water inside the pond: water constantly pushes against the walls. They must also be water tight (impermeable), so the pond does not leak.

The water inlet, located above the pond water level, is used to let water into the pond and is closed off after the pond is filled.

The drainage system is used to empty the water from the pond when the farmer is ready to harvest the fish.

There are many ways of making inlet and drainage systems: the most important criterion is that they work. But the walls are especially important: they are all that keep the fish inside the pond. The walls must be built carefully.

Pond construction follows the same principles whether the pond is a single backyard pond or part of a large fish hatchery. These are the steps in pond construction:

- Survey the land
- Mark out the area of the pond
- Measure and mark out the walls
- Excavate the pond bottom, if necessary
- Build the drainage system
- Build the water inlet
- Build the walls
- Seal the pond bottom and wails

Each of these steps will be discussed in detail in the following pages.

Survey the Land

The first step in the construction of a fish pond is marking the area of the proposed pond. If the site chosen is a natural slope, the first thing to be done is to find out where the main wall will be built. The main wall should be marked off at the lower end of the pond, where the pond will be the deepest and the slope the greatest. This is where the pond's drainage system will be put. If the pond is to be on a flat area, the pond bottom itself must be made with a slope so the pond will drain. This is done by digging one end deeper than the other end. Remember: the main wall is always at the deeper end.

DETERMINING THE SLOPE

Even flat ground usually has some kind of slope, although it may be very little and hard to see. So, before constructing the pond, the land is surveyed to find out which way the land slopes and what that slope is. There are a number of ways which can be used to determine slope. The way outlined here probably would not be used by many farmers If they were building a pond on their own, but this is an accurate method of determining slope and should be encouraged if at all possible.

To survey the land for slope, some stakes (long, straight pieces of wood), some string (fishline, etc.), and a carpenter's level are needed.

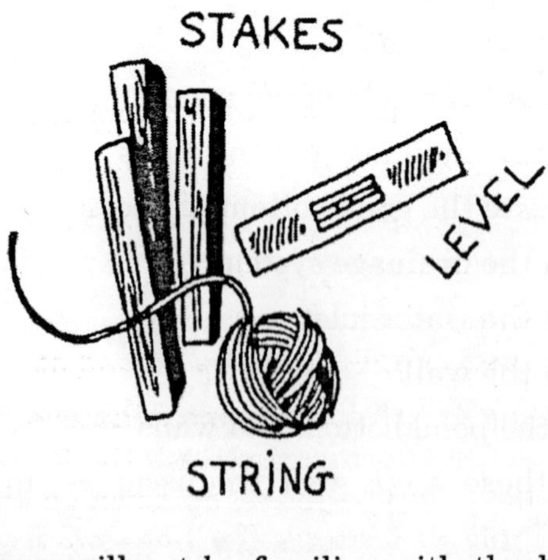

Most farmers will not be familiar with the level, a device that has an air bubble trapped inside which rests between two drawn lines. When the level is placed on the ground, it shows whether the area is flat or sloped: if it is straight or flat (level), the bubble stays in the middle between the lines; if the land slopes, the bubble will move to the right or left of the lines, depending upon the direction of the slope.

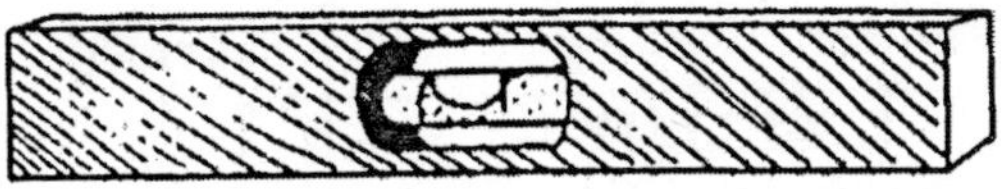

Carpenter's Level

Farmer's who cannot find a carpenter's level can make a level by getting a small light weight container. They should then place the container on a known horizontal surface, add water, and draw a line around the inside of the container at the water level. Then, if this container is placed on a slope, the water will shift away' from the line to show the slope.

When all the equipment is gathered, measure the slope.

- Look at the land and decide which part is higher.
- Drive a stake or a piece of wood or bamboo into the ground at the highest point.
- Walk away downhill from the stake about 100cm. Drive another stake into the ground at this point.
- Tie string or fish line or vine (whatever is being used) between the two stakes. Attach the level to the string. Then move the string up and down on the stakes until the bubble is between the lines on the level, or the water level is even with the line marked on the container. This will mean the string is level between the stakes, even though the stakes are in the ground at different heights.

- Measure the height of each string by measuring from ground level to the place where the string is tied.

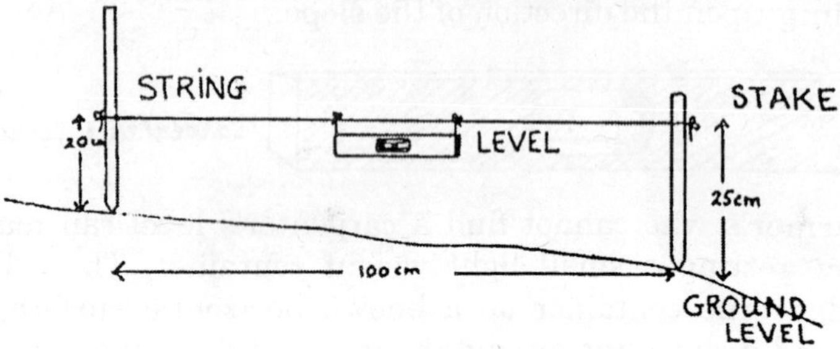

This drawing shows that one string is tied at 20cm; the other is tied at 25cm. Therefore, one end of the area is 5cm lower than the other. The distance covered by the string is 100cm, so the slope is 5% (over 100cm of ground, the elevation changed 5cm). Since a slope of 2-5% is good for a fish pond, this site has a satisfactory slope for a pond.

Other Ways of Determining Slope. As mentioned earlier, the above method of measuring slope Is a good one, but it may be difficult for some people to do. It is possible to calculate slope roughly. A farmer, who realizes that what he is looking for is a way to place his pond so that the water can enter from the water source and drain away well, can figure the slope of his land by doing such things as rolling a ball or other round object and watching carefully to note where and how quickly the ball rolls. A good slope would mean a slow-rolling ball. A variation of this involves throwing a quantity of water, or a mixture of water and dye on the ground and watching the path it takes and its speed as it moves along the ground.

It is important to consider slope carefully. A well-placed pond with good drainage is easier to care for and has more chance to be successful. It may be necessary for the pond owner to measure his land only once to find a good location. Or it may be necessary to repeat the measuring a number of times. This is probably a good thing to encourage since locations which look alike to the eye often have enough difference in slope to make a big difference to a fish pond. Also, determining slope is a larger project if more than one pond is being built. Then the ponds must be laid out in relation to each other.

There may be several areas which have the correct slope, but only one which is good in terms of getting the water into the pond from the water source and out of the pond easily. For example, the farmer might like to drain his pond so that the water irrigates his fields. Therefore, he will want to keep this in mind when he decides upon the exact placement of his pond. Likewise, if he is building a pond on a hillside in back of his house, the slope may be perfect, but he will need to avoid drainage into his buildings.

Once the slope is found, the location of the main wall can be determined. Of course, if the pond is built on flat ground, it will have four walls. If the pond is a barrage pond, it may only have one wall. The number of walls depends upon the land. The shape of the land may mean that one wall or two walls or four walls will be needed.

Mark out the Pond Site; Measure the Walls

Now that the slope is known, the place of the main wall is known. The main wall is at the end of the pond which will be deepest, and is the wall where the drainage system will go.

Mark out the main wall, and any other walls that will be built, with stakes. The walls, when finished, will be wide: it does not matter so much where the stakes are placed within the width of the planned walls, for they are to be used as height markers.

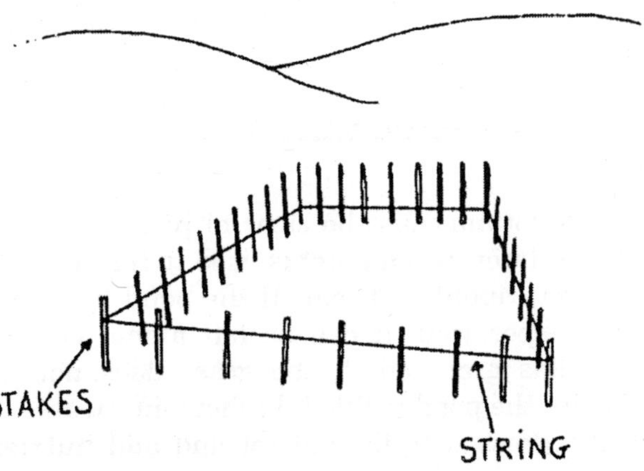

The farmer has to plan the depth of his pond and the height of his wall. If the pond is going to be 2m deep at the deepest end, for example, the walls should always be at least 30cm higher than the water level for a small pond, and at least 50cm higher for a large pond. Also, the walls will settle after they are finished, so it is best to make the wall 10% higher than the desired final height of the wall. A 2m deep pond, therefore, would have walls with a total height at the deepest point of 2.5 or 2.6m [height of wall before it settles = depth of pond + 30cm (for small pond) or 50cm (for large pond) + 10% of depth and 30 or 50cm].

Tie strings to the stakes along the main wall line, at a height of 2.5 or 2.6m for a pond whose deepest end will be 2m. Use a levelling device to connect strings to the stakes marking the other walls, if the pond has other walls, at the same level as the string marking the height of the main wall. The strings are the building markers. When the walls reach the strings, they are the right height.

Dig the Pond Bottom

As stated before, the pond bottom must slope downward from the shallow end to the deep end to help drainage. The pond bottom usually has a slope of from 2 to 5% (A slope of 2% would mean that for every 100cm change in length there is a 2cm change in height).

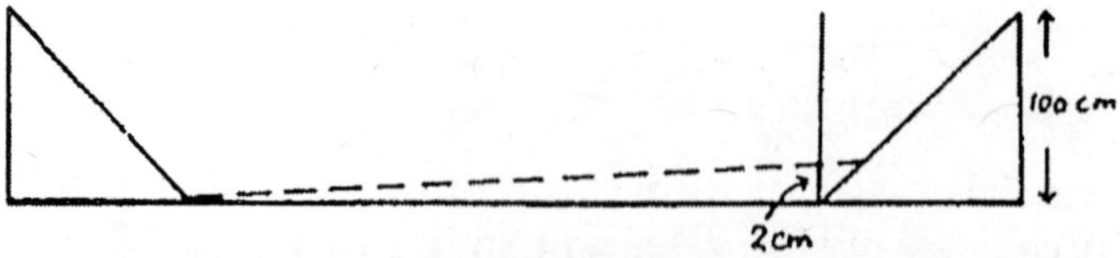

The pond bottom must be clear of rocks, roots, trees, and stumps so that later, when a net is used to harvest the fish, the net will not get caught and tear. If the pond bottom is already smooth and slopes well, it can be left alone. Or, if the pond bottom only has grass on it, the grass does not need to be removed before the pond is filled. In fact, once water is added to the pond, the grass will die and rot and add nutrients to the water.

If the pond bottom does not already slope downward, excavate (dig out) the bottom area of the pond until a good slope for drainage is made.

Adjust the height of the strings tied to the wall markers if digging the bottom has changed the height.

Keep the soil which was dug out of the pond: when the pond walls are finished, the soil can be placed on top and planted with grass. This fertile topsoil will root grass easily; this grass will help keep the walls from eroding (washing away).

The pond bottom can be excavated by hand or by using machines, like bulldozers, if they are available. Remember: if the land for the pond is chosen well with regard to the natural topography, only a small part of the pond bottom will need to be dug out. The most important thing is to have the pond bottom slope so that the pond can be drained.

Build a Drainage System

A drainage system is anything that is used to empty the pond. It consists of the outlet system for letting water out of the pond and the drainage ditches which carry the water from the pond away.

As stated before, the best and easiest way to have a good drainage system is to build the pond in a place which provides a good slope -- on a hill, for example. This is the first step. Then, there are many different drainage systems which can be put into the pond. Some of these drainage methods are expensive; others are very inexpensive.

The drainage system must be built before the pond walls because some drainage devices go through the walls (In some countries the drainage is done by knocking a hole in the wall of the pond. When the pond is dry and empty, the hole is patched up).

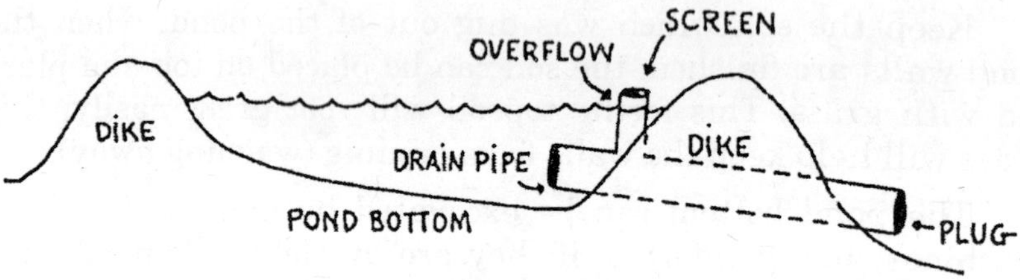

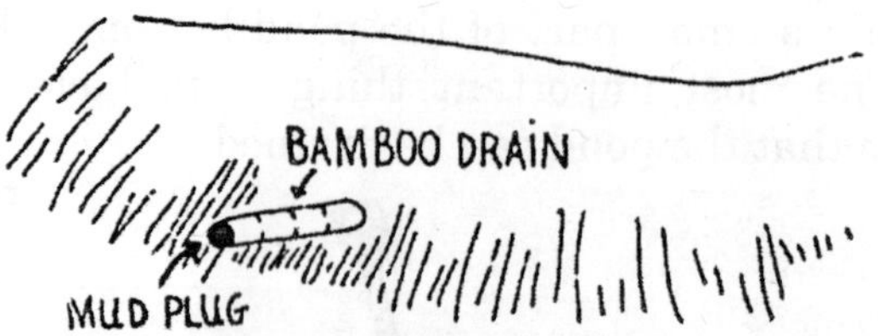

One of the easiest ways to drain the pond is to place a bamboo or plastic pipe through the base of the wall into the middle of the pond. The end of the pipe which is inside the pond has a screen over it to keep fish from entering the pipe. The other end of the pipe, the end that is outside the pond, is plugged with wood or clay. To drain the pond at harvest time, the plug is pulled out.

Two other methods of draining the pond which work but are not used as often, are the siphon and the pump. A siphon is merely a flexible plastic or rubber tube. One end of the tube is in the pond near the bottom; the other end is placed on the ground outside the pond. A vacuum is produced in the pipe by sucking at the end outside the pond until water begins to flow out. The end of the pipe inside the pond must be kept in the water or the siphon will not work.

The pump is usually not a good idea for a fanner because the engines that are used to run the pumps are costly and often not available, or gasoline to run them is costly, or they must be given frequent attention so they will not break down.

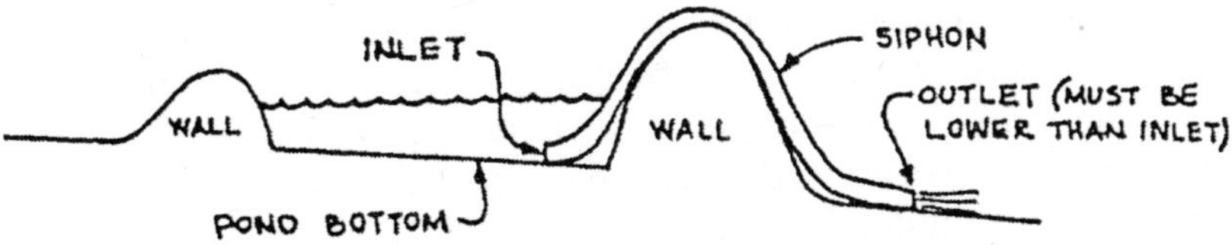

All ponds must be drained for harvesting fish. Also, it is a good idea to let a pond dry out completely once every year or so to get rid of any unwanted fish and/or disease-causing organisms.

The following are some tested, effective drainage systems a farmer can consider for his pond.

RIVALDI VALVE This valve was named after a farmer in Paraguay who first used the system. It is an easy and good method to use in a small fish pond. A farmer who is building only one small pond for family use would find this valve a good choice for his needs.

The Rivaldi valve is a flexible plastic pipe. Place the pipe on the ground before the wall is built. Build the wall. Then turn up and tie the pipe to a stake. Tie the pipe end at a level which is somewhat above the usual level of the water in the pond. Keep the pipe up and tied to the stake until it is time to drain the pond. Then, untie the pipe and let it lie on the floor of the pond until the water is out of the pond. At other times, the pipe works as an overflow to let out water after a heavy rain: when the water level in the pond reaches the top of the pipe, water will flow down the pipe and out of the pond.

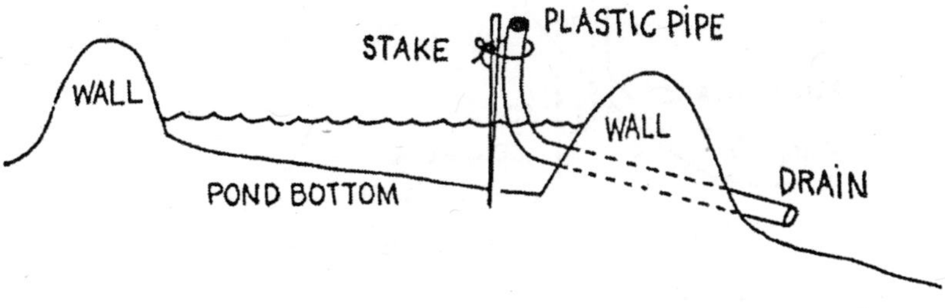

The Rivaldi valve should have a screen over the end inside the pond to keep fish from going out of the pond while the pond is being emptied or drained.

ELBOW JOINT A variation of the Rivaldi valve, this consists of two metal or plastic pipes connected by an elbow joint. The joint lets the upper pipe be turned down to drain the pond. The joint is screwed onto the ends of the two pipes, one of which extends under the wall and the other above the surface of the water. This drainage method is also called a "turn-down" pipe because it is actually turned on its side to drain the pond.

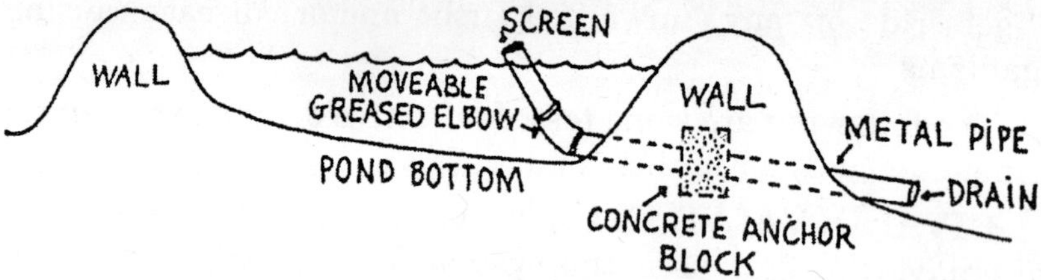

BOTTOM-WATER OVERFLOW This drain takes water directly from the bottom of the pond where oxygen levels are the lowest. The Rivaldi valve and elbow joint do this also, but each of these requires that the pipe be lowered so the pond can be drained. The bottom-water overflow regulates the depth of water without any need for moving the pipes. When new water is added to the pond, the less-oxygenated water at the bottom drains out automatically.

This type of drain is relatively complicated and usually difficult to build. For a small fish farm operation, it would probably not be worth the effort.

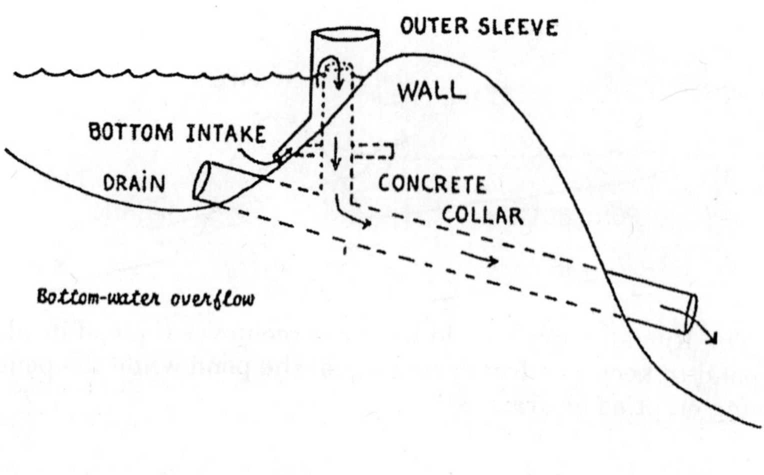

Bottom-water overflow

DOUBLE SLEEVE OVERFLOW This drainage system is built like the turn-down pipe, except a large pipe is placed over the section of pipe which extends above the pond's surface. This outer pipe should be longer and wider than the inner pipe, which is placed so that it is about equal in height to the depth of water desired in the pond.

When fresh water is required in the pond quickly because the water Is too warm for the fish or because the oxygen levels are low, all the farmer has to do is to add water to the pond. The double-sleeve overflow automatically drains the stale water from the bottom of the pond.

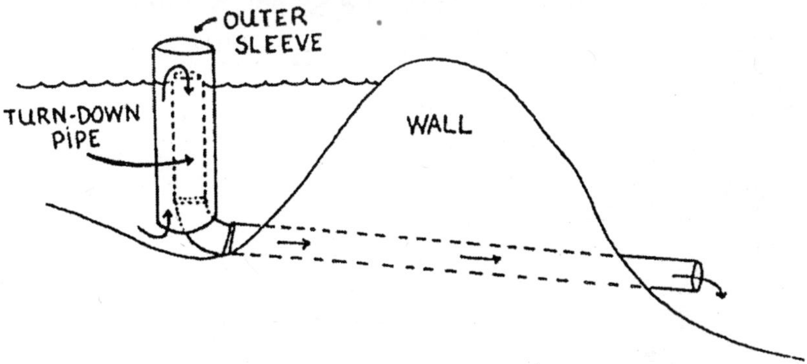

SLUICE A sluice can function in a number of ways in a pond. It can be a screened gate in a water channel going into the pond, or a drainage gate leading water out of the pond.

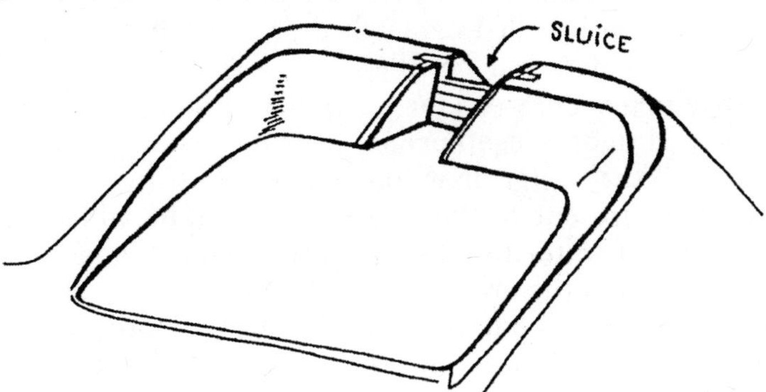

In a pond, a drainage sluice gate is anchored into the main wall by extending the sides of the sluice into the wall so the sluice structure stands upright. The sluice is constructed at the center of the main wall before the dike is built.

The sluice can be made of wood, cement, or brick. It can have one or two wooden gates which are removed to empty or fill the pond. A sluice also can have a screen gate to keep unwanted fish from entering at an inlet and pond fish from leaving at the outlet.

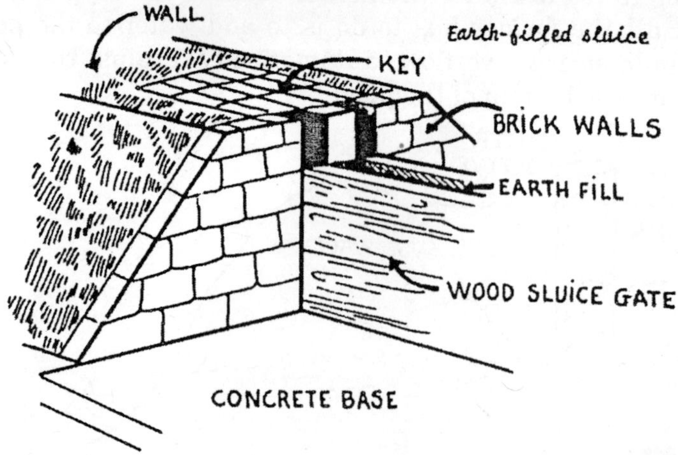

IMPORTANT: The wooden gates of the sluice must fit into the slots well, but easily. The wood will swell to make a tighter seal as it is soaked by the water in the pond. The slots (grooves) can be filled with several strong, long, narrow boards which have been bevelled or notched so that they fit together tightly. Or the slots can be filled with single pieces of wood. When single pieces of wood (or a number of boards which have been fastened tightly together) are used in a sluice, the pond is drained and the water flow regulated by lifting the entire wooden structure out of the groove to a height which allows some or a lot of water to flow out of the pond. When separate boards are used in the grooves, the boards are taken out one at a time. If a small flow out of the pond is desired, only one board may be taken out. To drain the pond, all the boards are removed. In a sluice having two wooden gates, the space between the gates can be packed tightly with earth. This will help seal the water into the pond.

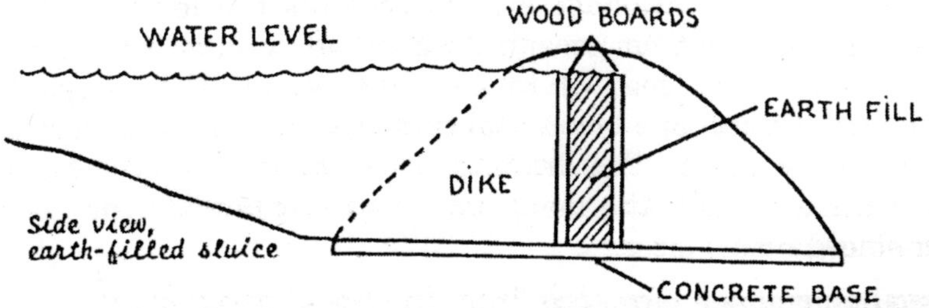

MONK The monk is very much like the sluice, but it is not built into the pond wall the way the sluice is. Sometimes the back of the monk does touch the wall, but it is not built into the wall. Also, a monk is never used at the inlet as a sluice can be.

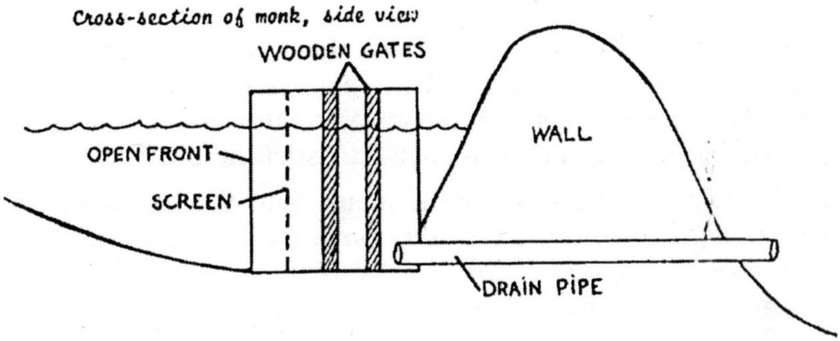

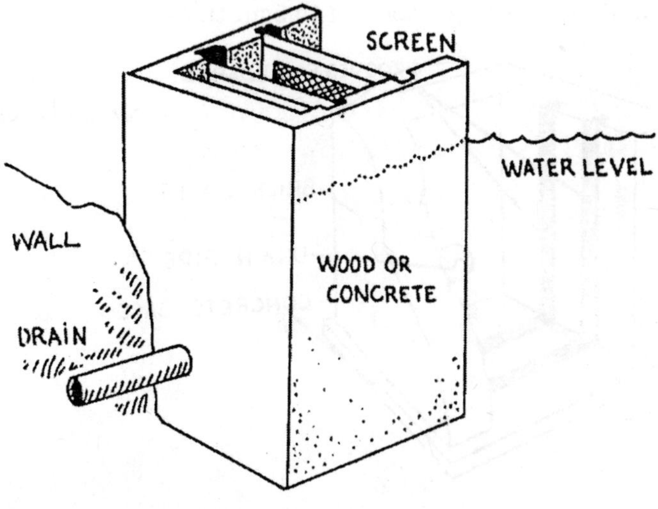

A monk-type drainage system controls the level of water and prevents fish from escaping when the pond is being filled. It also allows for good drainage of the pond. The completed structure consists of a horizontal drainage pipe and the vertical structure, or monk. The drainage pipe must be placed before the walls are built; the monk may be built outside the pond, and placed inside later.

The drainage pipe runs from the back of the monk under the pond wall. It should be between 20 and 40cm in diameter; if piping of this diameter is not available, two pipes may be used. For good drainage, place the pipes 30 to 40cm lower than the pond bottom. Make sure the drainage pipe is on solid ground so that the pipes do not bend. Bent pipes are difficult to clean out when clogged.

The monk itself is a structure which is closed on three sides and open in the front. The open side should face the inside of the pond and should be at least 30cm wide; the entire monk should be at least 40cm above the surface of the water.

The two parallel sides of the monk, and the bottom, have grooves cut in them: a monk may have two or three grooves. One groove, or part of a groove is always for the screen. The other groove(s) is for the boards.

Monks can be made of wood, concrete or brick. A wooden monk should use strong wood -- 4 to 5cm thick.

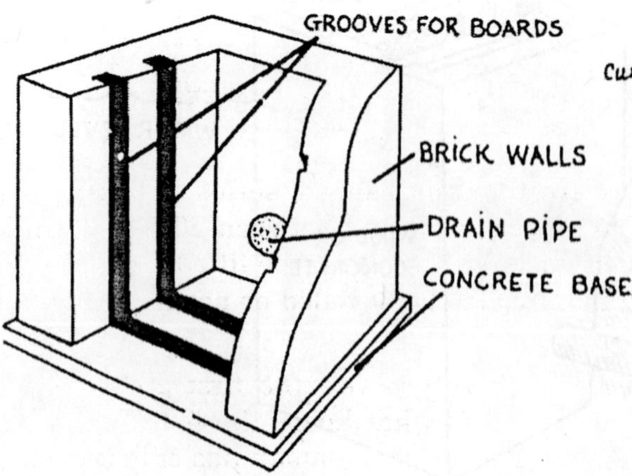

Cutaway of monk

A concrete monk should be reinforced with metal. Before the concrete is poured, a wooden form shaped like the monk is made and oiled. A frame, slightly smaller than the wooden form, is made of chicken wire, or some other strong wire, and set down inside of the wooden form. The concrete is then poured into the form. A good concrete mixture for monks is 1 part cement, 2 parts clean sand, and 4 parts crushed stone, by volume.

If the monk is made of concrete, the grooves can be shaped by bending iron rods into a "U" shape. Remember, the grooves must be sunk into the sides and bottom of the monk.

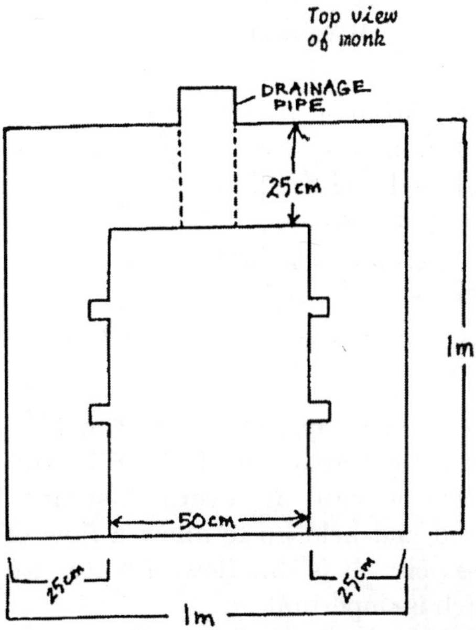

The grooves are filled by using a series of boards -- wide enough to fit the grooves well and between 20 and 30cm high. Each board has a hook on it so it can be lifted from the groove easily; the boards may also be bevelled or notched so that they fit together well.

If the monk has three grooves, the first groove can be a large screen. The screen is what keeps the fish from escaping as the pond drains. However, if the monk has only two grooves, a smaller screen can be placed above or below the boards in the first groove. Placing the screen at the bottom allows water to drain out from the bottom of the pond.

HERRGUTH MONK This is a monk with three grooves. A large screen is in the first groove. The large screen is better than a smaller one because it does not get clogged up as easily as a small screen.

The second groove holds a series of boards. The lowest board can be a small screen. Water flows through the large screen in the first groove, and through the small screen in the bottom of the first series of boards, up and over the third series of boards into the drainage system.

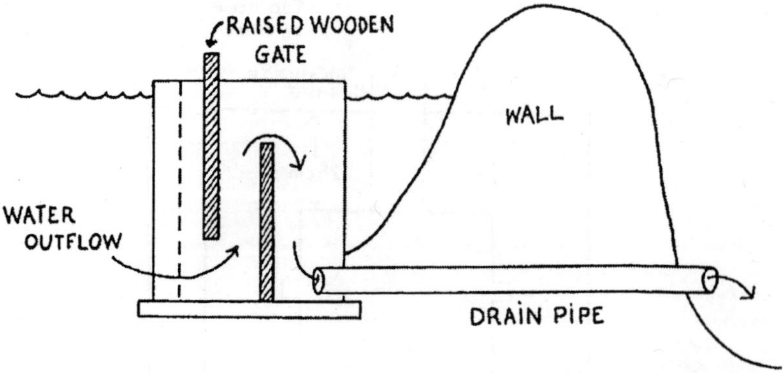

There are other ways this kind of monk can be built. For example, the second groove could be filled by a large wooden gate (one piece of wood or several fastened together) which could be raised and held up to allow a flow of water from the bottom of the pond. It is this flow of water from the bottom of the pond which is important.

The Herrguth monk would probably not be used in a pond which is filled by rainwater. In these ponds -- sky ponds -- a regular monk is used, and the space between the two wooden gates is packed with mud to make a watertight seal which lasts for the fish-growing season and is removed when the pond is drained for harvest.

SOME NOTES ABOUT MONKS Be careful with screens. Bamboo slats can be used instead of screening if the fish are large. But for fry, the holes should be less than 2mm in diameter. Often the screens are made by poking small holes in sheet metal. The screen mesh can get larger as the fish grow. A valve is sometimes placed on the drainage pipe behind the upright part of the monk. This is used to control the draining speed and is easier to do than to move the boards in the grooves. A large catching ditch can be made in front of the monk to help with taking fish out of the pond when the pond is being drained for harvest.

DRAINAGE DITCHES Drainage ditches are channels which should be dug on the bottom of the pond to help the water flow out. Lining the ditches with stones helps the water flow. A small family pond does not require this system of drains. The only real requirement for drainage is a gentle slope.

This is, the time to build other ditches which may be needed. For example, if the farmer wants to use the water from his fish pond to irrigate his land, he will want to construct the ditches or chann3ls which will carry the water from the pond to the field or to storage tanks for use later. Therefore, the farmer must consider carefully where the water which is draining from a pond is going to go. If the pond is being fully drained, and the pond is built on flat ground, he should build drainage ditches around the outside of his pond to drain the water away from the walls. These ditches should be 30-40cm deep.

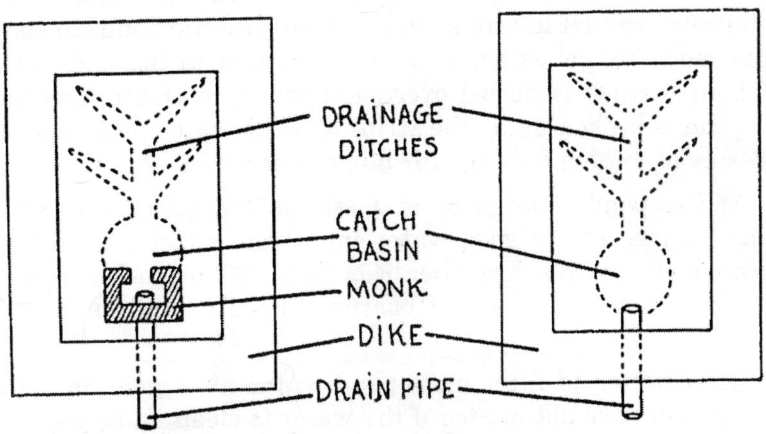

Water Inlet

All ponds, except for those filled directly by a spring or by rainwater, need water inlets. The water inlet must be constructed so that it supplies adequate quantities and quality of water, and so that it does not allow unwanted fish or other materials to enter the pond. This usually means there must be a channel of some kind to bring the water to the pond from the source and a filter of some kind to keep the water which goes into the pond clean and free from predators.

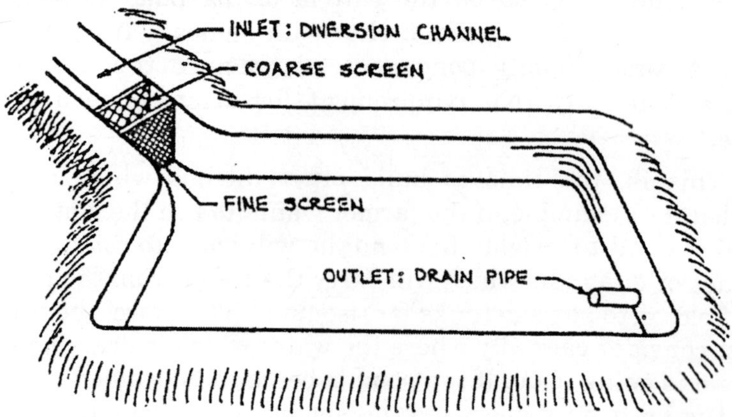

A water inlet can be as simple as a bamboo pipe of good diameter running from a water source through the wall into the pond. Remember: the inlet pipe should be placed above the water level so that incoming water drops into the pond. In some areas, such things as bamboo strips are tied to the end of the inlet pipe which is placed over the pond. The water flow into the pond is broken up by the strips and the water picks up and takes more oxygen from the air into the pond water.

If the pond is large or is a stream fed barrage pond, a sluice makes a very good water inlet. The sluice can be one piece which controls flow when it is lifted to various heights, or the sluice can be a series of boards slipped in and out of the grooves.

It is better to filter most pond water as it goes into the pond. Filters are not needed if the water is clean and clear and the farmer knows the source is free from unwanted fish. But if the water is muddy, or has lots of leaves or debris in it, a filter helps keep the water quality good.

A filter can be placed at the beginning, middle or end of the channel which brings water to the pond. Usually filters work best near the water inlet. Filters can be made very simply. Remember they must keep unwanted fish out and pond fish in.

A wire screen makes a good filter. The picture above shows a sluice with a gate with fine screening to strain incoming water of pieces of debris and other unwanted fish and materials. Note the screen fits into the water channel exactly.

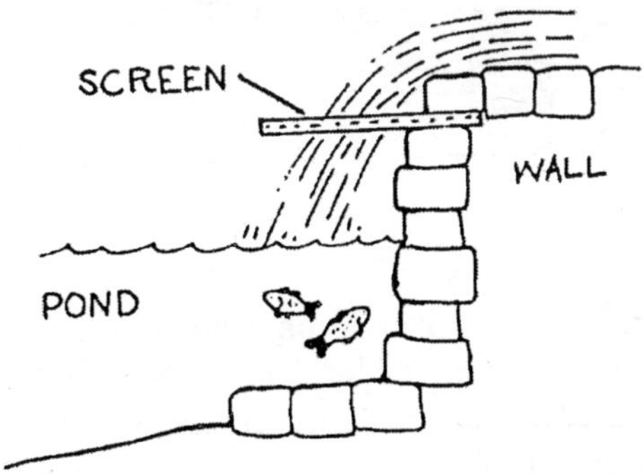

The horizontal screen at the left is very effective. Here the screen is placed so that the water passes through as it falls into the pond. This screen merely juts out from the wall at the inlet.

In the version below the horizontal screen has a vertical screen wall attached to it. This short wall prevents fish from going over the screen.

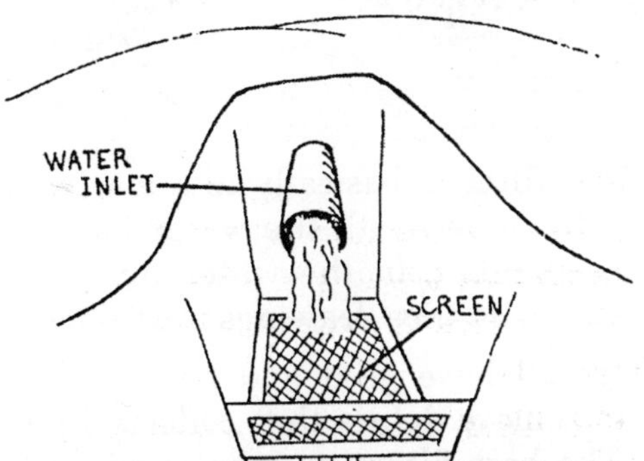

In any variations of these kinds of filters, the screens should be assembled into one piece for easy removal as a unit for cleaning. There are other ways of filtering the water:

A nylon mesh bag makes a good filter, as long as it is partially submerged in the pond so that it does not tear as the bag catches fish or other material from the water source. Check it periodically.

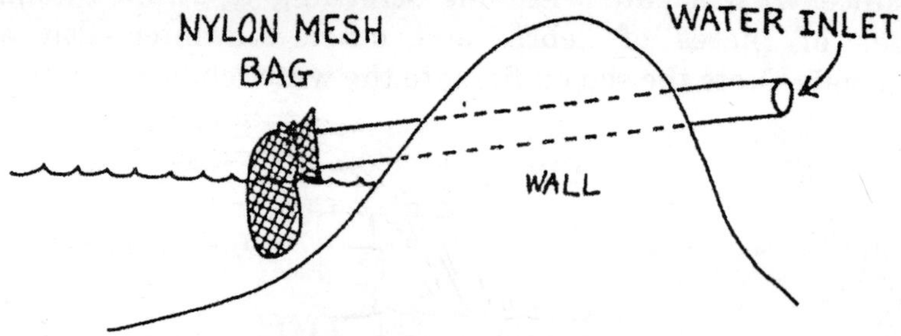

A sand and gravel filter is particularly useful for cleaning out fish and eggs. It requires building a smaller pond or tank at the water inlet. If a filter is built in the earth it must be lined with a waterproof liner.

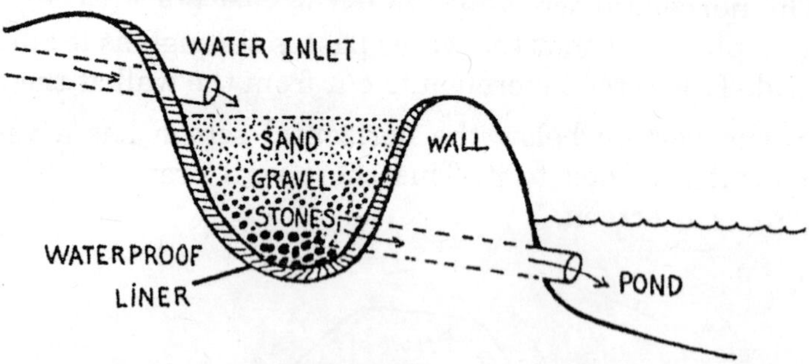

A saran fiber filter is basically like a wire screen that is placed horizontally underneath the water inlet. However, it is placed in a box standing in the water and uses saran fiber material instead of wire (See drawings next page).

These filters all have good and bad points. All must be cleaned often to remove debris that collects in them from the water source. The best filters are the sand and gravel filter, and the saran filter, but these are more costly than the others.

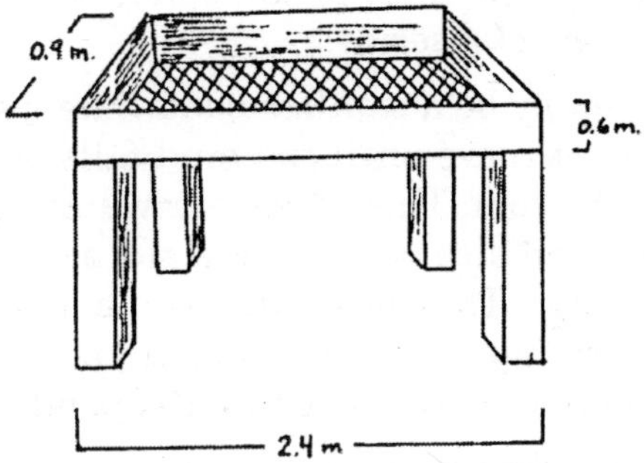

Saran fiber filter

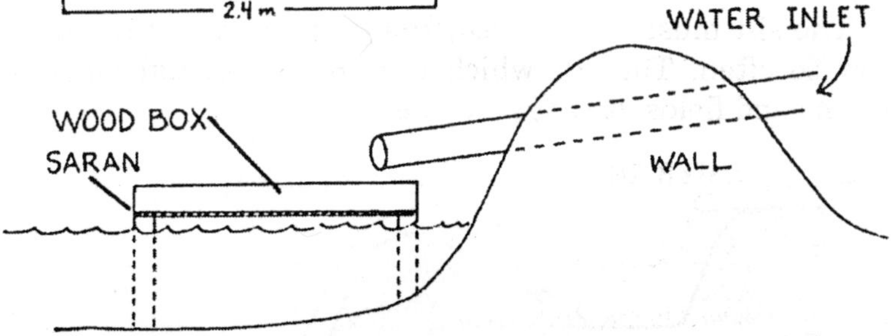

The farmer should examine his water source carefully before deciding on the kind of filter. If the water is very muddy, or has lots of leaves and grass in it (organic matter), he can use the wire screen. If the water source is free of organic material, the mesh bag will work because it is not likely to be torn. If the water contains unwanted fish and eggs, as well as a lot of organic matter, the saran filter or the sand and gravel filter is best.

To clean the filters, remove them and clean them with a brush and fresh water. Or flush the filter with water in the opposite direction of the normal water flow. This is called backwashing.

IMPORTANT: Filters must be kept clean to be of any use. These filters should be cleaned each time water is let into the pond.

SILTATION TANK One other structure which should be built at the water inlet, when necessary, is called a siltation tank. Silt is the mud that is suspended (floating) in water. Silt can become a problem when it clogs the gills of the pond fish so they cannot breathe. If the water source has a lot of mud init,a siltation tank should be built at the inlet to the pond, or at the inlet to the first pond, if it is one of a series.

The siltation tank can simply be a smaller pond. The water flows into this pond and is kept there until the mud falls out of the water and settles on the bottom. Then the clear water is :et into the fish pond. Siltation could also be done in a storage tank made out of old oil drums, etc. The important thing is that something be constructed or set up so that the silt has a chance to fall out of the water before that water goes into the pond.

The silt must be removed from the siltation tank or pond every so often. The silt which is removed should be used in gardens and fields: it is very fertile.

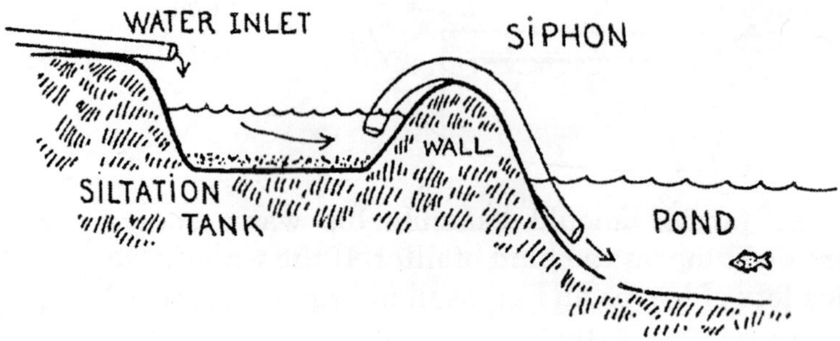

Build the Walls

The walls (dikes, dams, levees) have to withstand the pressure of all water in the pond. the They also have to be watertight to keep the water inside the pond.

The construction of the walls depends upon the kind of soil in which the pond is being built.

A soil which is a mixture of sand and clay is best. If pure clay is to be used, it must be mixed with other soil before it can be used. Pure clay will crack and leak. Do not use turf, humus, or peaty earth. All stones, pieces of wood, and other materials which might rot or otherwise weaken the wall must be removed before building begins. If the soil contains enough clay, the walls can be built by placing layers of soil 20cm deep over the drainage pipes and tamping each layer down until it is compact.

The finished height of the wall should be about 30cm above water for small ponds and 50cm above water for large ponds. The width of the wall at the top should be about equal to its height. For a large pond, the wall is never less than lm wide at the top; most walls are built so that two people can walk side by side along the top.

Tamp the soil down with a simple tamping tool. Some people use a large rock or even their own weight by Jumping up and down on the soil. The important thing is that the soil must be packed down very tightly.

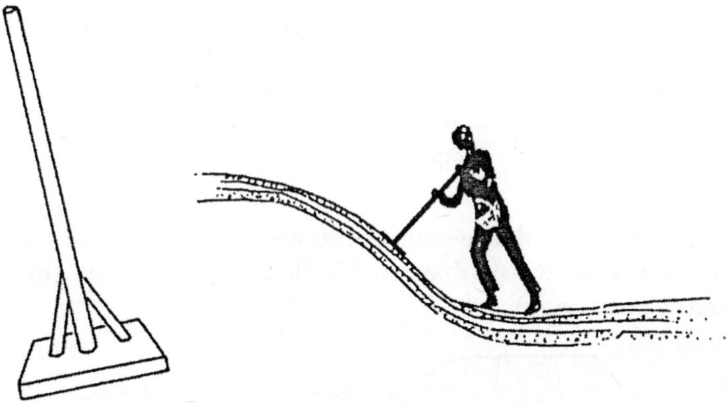

One way to build pond walls in soil that does not have a lot of clay or is very sandy is to build a "key." The key is made of clay soil (it can be pure clay) and adds strength to the walls. To make a key, dig a trench (or shallow hole) about 1m deep and 1m wide in the center of the places where the walls will be. Then bring clay soil and pack it tightly into the trench. Also put a thick layer of clay soil on the pond bottom and pack that down tightly. The clay layer on the bottom and the key run together as shown. This connection of the bottom and the key helps prevent leaking. The drainage pipe should be placed in the clay lining.

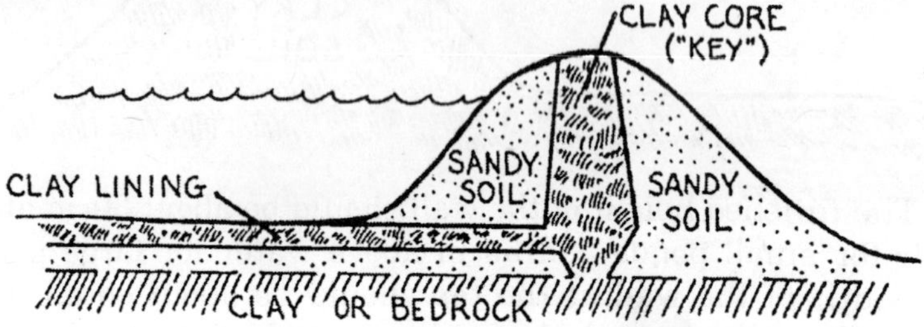

If the farmer has a soil which is a mixture of clay and sand, and he is not sure it is strong enough, he may still wish to build a clay key. Or he can build a key using the same soil used in the wall. This key must be packed down very tightly.

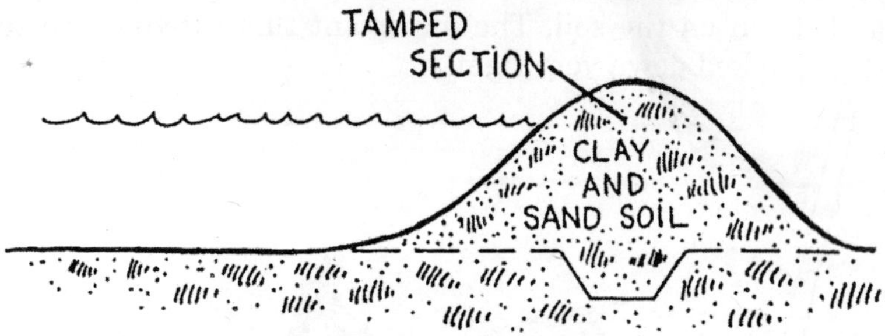

The type of soil determines the ways in which the pond can be prepared so water does not leak out (see "Seal the Pond Bottom", next page).

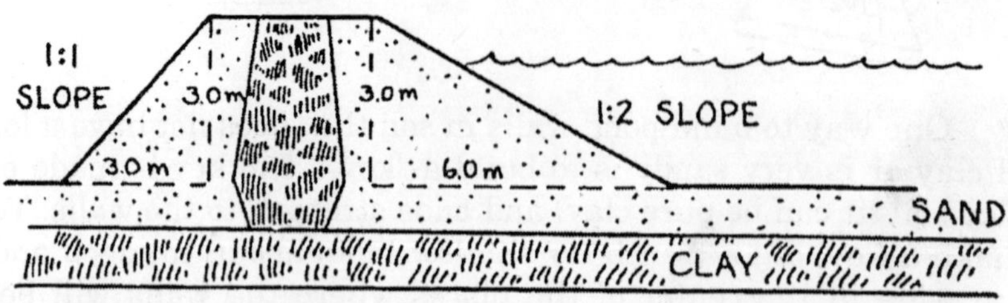

The soil also determines the slope of the walls. Soil with a lot of clay in it can have a greater slope on the outside wall than on the Inside wall. A typical wall is built with an outside slope of 1:1 and an inside slope of 1:2. A slope of 1:2 means that for every change In length of 2m there is a change of 1m in height.

Once the walls are constructed, the farmer should plant grass on them. The grass roots help to hold the wall together and prevent erosion of the soil. However, NEVER plant trees on the wall. As the tree roots grow they will crack and destroy the wall.

Seal the Pond Bottom

The last step in pond construction is sealing the pond bottom so that it does not leak. If the soil has a lot of clay in it, no special sealing is needed. If the bottom is sand or gravel, it should be sealed to help it hold water. One way to seal the pond is to build a clay core into the wall and extend the clay over the bottom of the pond as a lining. This kind of sealing must be done when the walls are built. After the walls are built, there are other methods you can use for sealing the pond.

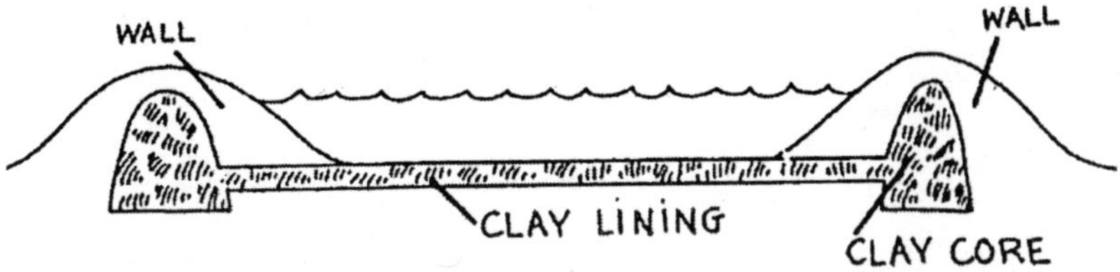

A pond can be sealed using hollow cement blocks, but this is expensive. Another method of sealing the bottom calls for using a sheet liner made of polyethylene plastic, or a rubber liner. The waterproof sheet is placed on the pond bottom and around the sides in one piece (the farmer may have to tightly seal several sections together), then covered with soil.

Another technique, recently developed in the USSR, is called a "gley" or "biological plastic." "Gley" can be made in the pond in this way:

- Clear the pond bottom of debris, rocks, and all other materials.
- Cover the pond bottom and sides completely with animal manure. Apply the manure in an even layer.

- Cover the animal layer with banana leaves, cut grasses, or any vegetable matter. Make sure all the manure is covered.
- Put a layer of soil on top of the vegetable layer.
- Tamp the layers down very well.
- Wait 2 to 3 weeks before filling the pond.

5 Preparing the Pond

The last pages of the construction section presented several ways of sealing the bottom of the pond so it will hold water better. This section tells what has to be done to prepare the completed pond for the fish.

Conditioning the Pond

If the pond is an old one from which the fish have been harvested, plow it completely. Plowing turns the ground over so that it dries well. Clear the bottom of any twigs, stumps, branches, or dead fish. Any predators (snakes, frogs, etc.) must be taken out by hand or poisoned (see 'Problems of Fish in Pond' for more information on this subject). Then smooth the bottom out again. When the pond is dried enough, the soil will have large cracks in it.

After the pond is plowed, cleared, and smoothed, it should be conditioned with lime.

Whether the pond is old or new, a layer of lime should be placed on the bottom of the pond. Place the lime on the pond two weeks before the water is put into the pond.

Lime conditions the soil of the pond. It is not a fertilizer, but it helps fertilizers work. It is especially important to use lime if the soil has acids in it which might harm the fish. Lime can control these acids so they are not a danger. A farmer who is not sure whether the soil of his new fish pond has acids in it -- because he had no place to get his soil tested, or because he has never farmed the land -- is always safer if he puts lime on the bottom of the pond.

Lime comes in several forms: ground limestone; agricultural lime; hydrated (builders') lime; or quicklime. Of all these types, hydrated lime is cheapest to use because it is more concentrated.

Quicklime must be used carefully: it can burn if it touches the skin and is harmful if breathed Into the body. Farmers should be warned to use quicklime only with extreme care.

Lime should be put on the pond bottom at the following rates for a new pond:

Ground Limestone	1140 kg per hectare
Agricultural Lime	2270kg per hectare
Hydrated Lime	114kg per hectare
Quick lime	200kg per hectare

A word about limestone: In many areas of the world, limestone can be found locally. It is a soft stone and may be ground by the farmers themselves. It is a good idea to let farmers know whether or not limestone is available locally and to help them identify it if they can not already do so.

Filling the Pond

After the lime has been on the pond bottom for at least two weeks, let the water in slowly. The water should fall from the water inlet into the pond below, so that the water mixes with oxygen from the air as it falls into the pond.

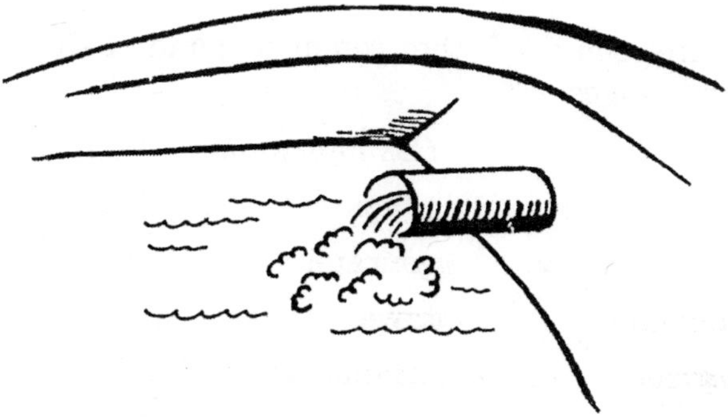

The water should not go in too quickly. If the water goes in too fast. the pond bottom will get stirred up and make the water muddy.

Let the pond sit for a few days after it has been filled. Then check the quality of the water in the pond -- before adding the fish.

Fish growth depends greatly on the quality of the water used in the pond. And the quality of the water depends upon where it comes from and what kind of soil it travels over. Testing the water quality means making sure that all the factors which relate to water are right for the fish. These factors are: temperature, oxygen content, pH, turbidity, hardness, alkalinity, and nutrient availability (source of food for the fish). The farmer does not need to know these particular words to raise fish well, but he does require a working knowledge of the factors that are part of the water world in which the fish live.

TEMPERATURE

Fish are cold-blooded animals; that is, their body temperatures depend upon the temperature of the water in which they live. Every fish species has a temperature range within which it grows quickly. This is called the optimum temperature range, and it means that this fish grows best at temperatures within that range. In a fish pond, the fish should live at their optimum to grow well. However, since fish have different temperature requirements, the farmer must choose the fish which will grow best in the temperature range of his pond.

Here are some of the common pond fish and their temperature ranges:

Genus, species.	*Common name*	*Temperature °C*
Tilapia mossambica	tilapia	25-35
Osphronemus goramy	gourami	24-28
Puntius javanicus	tawes	25-33
Cyprinus carpio	common carp	20-25
Ctenopharyngodon idellus	grass carp	25-30
Anguilla japonica	eel	20-28

This chart shows that all the fish on this list could live in water that is 25°C (77°F). The chart also shows that an eel can live and grow well at 20°C, but that the tilapia and the grass carp will not do well at 20°C because this temperature is below the range in which they are comfortable. When the temperature goes higher or lower than this optimum, fish will not grow. Eventually, if the temperature goes too high or too low, the fish will die.

The farmer must watch the temperature in the pond water carefully, if the weather becomes unusually hot or cold. If it is possible, it is a good idea for a farmer to use a thermometer to find the temperature of his pond water. This can be done by using a thermometer which is used for taking temperatures when people are sick. The most important step is to guide the farmer to stock fish which will do well in the normal temperature ranges of his area. Then the temperature of the water will not generally be a problem, except in cases of unusual weather.

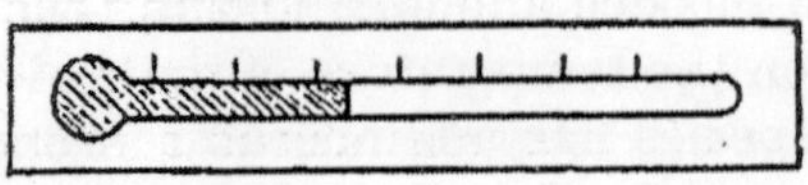

Some experienced fish growers can judge the water temperature by putting their arms in the water. Most people cannot tell temperature this way. But if the right kind of fish has been chosen for the pond, the farmer need only watch the fish to be able to judge the temperature of the pond' water. If the water is becoming too hot, the fish will not eat and will move very slowly.

If the farmer sees this behavior in his fish pond, he can take out some of the pond water and put in new, cooler water. Another way of protecting the water from getting too hot is to find a way to shade the pond, so that the sun does not shine directly on the water. The shading should be temporary because sunlight is important to the success of the pond.

The picture on the previous page shows a fish pond being shaded by palm tree branches stuck into the ground around the edges of the pond. As soon as the temperature of the water goes down, the branches are removed.

Temperature, however, usually does not act alone. If the fish are showing signs of distress because of hot weather; it is often a problem caused by high temperatures and low oxygen content.

OXYGEN

The farmer cannot see oxygen, so it may be hard for him to realize its importance. But it is worth taking the time to help a pond owner understand oxygen as a critical factor In the success of his fish pond. Oxygen lack is a problem which can occur at any time during fish pond operation, and there is a good chance the farmer will have to depend only upon his own knowledge of the problem and its cause to solve it immediately.

Fish, like all animals and human beings, need oxygen to breathe and, therefore, to live. Through a process called respiration, fish and human beings take in oxygen and give off carbon dioxide. Fish will not grow well when the oxygen supply is low; and if the oxygen level gets too low, they will die.

Oxygen is a gas. Human beings get the oxygen they need from the air. They cannot see it, or smell it, but without it they would die. Most fish can only get oxygen from the water in the fish pond. The farmer cannot see the oxygen in the water either, but he should realize that it must be there in sufficient quantity for the fish to live.

Oxygen troubles arise in a pond when the supply of oxygen is used up faster than oxygen is put into the pond. This happens to human beings too -- if too many people are shut into a room with no windows or air holes, the respiration of all these people uses up the oxygen. Soon, there is too much carbon dioxide in the air. The people have trouble breathing until a window is opened and fresh air containing oxygen is let in.

This is exactly what happens to fish in the fish pond. The fish are shut up in the pond, and if there is not enough oxygen entering the pond, they will have trouble breathing. And, if the problem continues, they will die.

Water contains tiny plants and animals called plankton. Most plankton are so very small that they cannot be seen without using a microscope.

The plants are phytoplankton:

Dictyosphaerium pulchellum

The animals are zooplankton:

Alona sp.

Water also contains higher orders of vegetation. These plants are much larger than the phytoplankton.

Azolla pinnata

Hydrilla verticillata
A favorite food of the gourami.

The fish and the zooplankton use oxygen and give off carbon dioxide in respiration; the phytoplankton and higher plants use carbon dioxide and sunlight to produce oxygen during a process called photosynthesis.

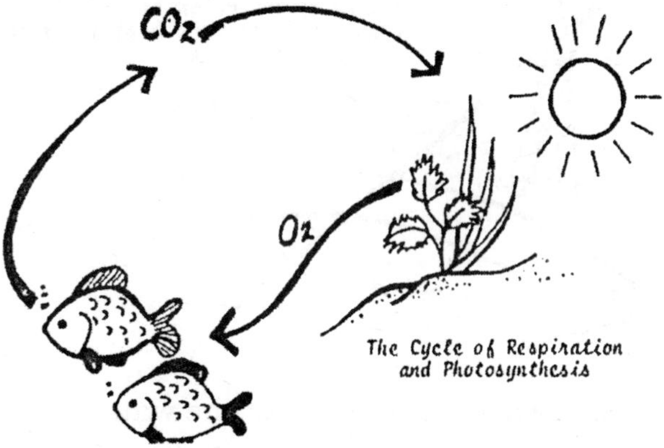

The Cycle of Respiration and Photosynthesis

The oxygen in a fish pond also is used up by the process of decay. Dead organic matter -- leaves, fish, other plant and animal material present in the pond, use up oxygen in the decay process called oxidation. Oxidation and respiration go on both day and night, while photosynthesis can take place only during sunlight hours.

Therefore, there are times during the day when the oxygen levels in the pond can be very low, and oxygen may have to be added to the water. Oxygen can be added to the pond water by taking out some of the old water which is low in oxygen and adding new water.

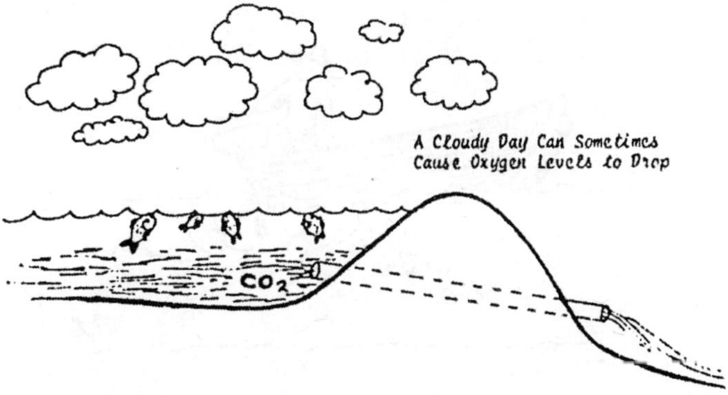

A Cloudy Day Can Sometimes Cause Oxygen Levels to Drop

The new water should be sprayed or bubbled into the pond so that the water picks up oxygen from the air as it falls into the pond.

Oxygen also can be added to pond water by:

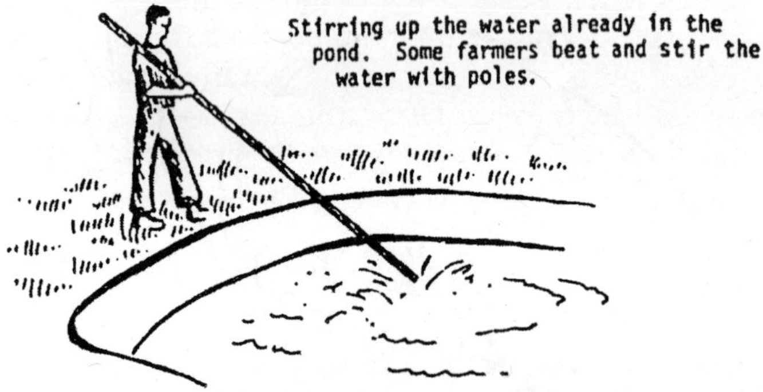

Some pond owners use oars to stir the water.

Other owners run small motors to bubble the water in the pond.

In addition, winds which are strong enough to ripple the surface of the water in the pond help the air and water to mix. Remember: any disturbance of the water made by man or by nature helps put oxygen into (aerates) the water.

Life under the water is a new idea to many farmers. And it is sometimes difficult to understand that the balances which exist on land are also present in the water. Oxygen is produced and used both above and below the surface of the water. The fish pond does well only when oxygen production and oxygen use are in balanced relationship.

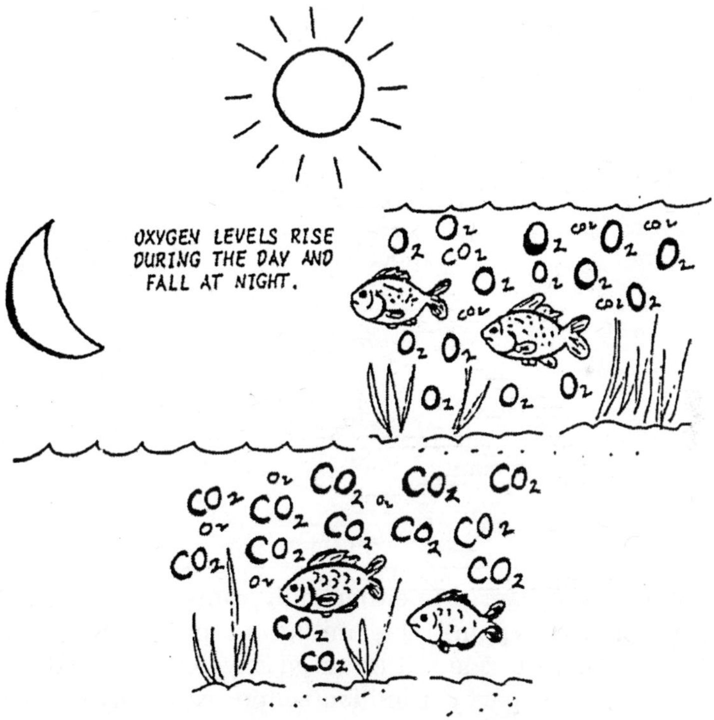

If the farmer understands the balance -- how oxygen is added and how it is used up, he will know how to watch for trouble before it happens. For example, if the color of the water changes from green to clear -- in a few hours or a day -- the phytoplankton are not producing enough oxygen. If the fish are at the surface of the water and seem to be gulping air, they may need oxygen. A Early in the morning, before the sun comes up, or a long period of no sunshine can be bad times because the phytoplankton need the sun to produce oxygen. Long periods of hot weather can create oxygen problems because the pond water gets warmer, and warm water cannot hold as much oxygen as cool water can.

The following table shows the difference in oxygen levels at various points in the day. For example, at 6 am, the temperature has remained steady, but the dissolved oxygen level has dropped to 6.3 mg. At 6 pm, after a sunny day, the dissolved oxygen level is 16.3 mg.

This table also shows that on a typical day a pond's temperature does not vary greatly. This illustrates why oxygen as a separate factor is much more important than is temperature.

MEASURED OXYGEN CONTENT COMPARED WITH TEMPERATURE IN ONE POND

Time of Day	Temperature °C	Dissolved Oxygen, mg/l
2 am	29	9.8
6 am	29	6.3
10 am	29	6.7
2 pm	30	9.4
6 pm	29	16.3
10 pm	29	10.7

Oxygen is measured in either milligrams per liter (mg/l) or parts per million (ppm). One milligram per liter of oxygen means that there is one milligram of oxygen dissolved in one liter of water. One part per million is approximately equivalent to one milligram per liter.

Fish begin to be stressed when the oxygen level falls below 4mg/l. For best growth, the oxygen levels should be above 5mg/l, but not more than 15mg/l. Above this level of oxygen, supersaturation often results (too much oxygen).

Sometimes, if there is a lot of sun and a lot of wind activity at the same time, and if the temperature is low, the water can become supersaturated with oxygen. Supersaturated water contains more oxygen than water can normally hold at a given temperature; it is a temporary condition. This condition can place stress on the fish. However, it does not happen very often in small ponds because the wind is not usually able to aerate pond water as thoroughly as it can in a large pond.

To determine the exact oxygen content of a pond, certain chemicals and equipment are needed. Dissolved oxygen is usually determined in the laboratory by the Winkler Method. Now, however, there are field kits available (Hach, LaMotte). These kits are, however, expensive, and certainly will not be available to most farmers.

A field kit for Testing Pond Water Costs about $ 30 (U.S.)

pH, HARDNESS, AND ALKALINITY

These three factors are not the same thing -- each one is a measure of a certain characteristic or characteristics of the water in a fish pond. Each of these factors can be measured exactly if samples of pond water can be taken to a laboratory to be analyzed, or if chemicals are available for testing the water in the field. Certainly if such testing is possible, it should be done.

However, many pond owners are not able to get their water tested and they do not have the right chemicals and equipment to do the tests themselves. For these people, it is best to stress the importance of using lime in their ponds. Lime is the proper treatment to correct imbalances in these factors, each of which is discussed in some detail here.

pH. pH is the measure of hydrogen ions (H^+ the water and is measured on a scale of 1 to 14. If the pH is between 0 and 7, the water is considered to be acid. If the pH is at 7, the water is neutral (not acid or basic). A pH of 7 to 14 means the water is basic. Fish grow best in a pH of between 6.5 and 9.0. Fish are very sensitive to low pH, or, in other words, to water which is acid. Most pond fish will die if the pH falls below 4 for a very long period of time.

Sometimes the pH of a pond can change quickly. For example, a heavy rain may carry acid from the soil in the dikes into the pond water. The best way to get the pH back to neutral is to add limestone (calcium carbonate) to the water by spreading it on the pond bottom or on the surface of the water. A fish like tilapia can tolerate pH from 3.7 to 10.5, but below a pH of 5, they are stressed and they will not eat.

Some people measure pH by tasting the water. If the water tastes sour or salty, it has too much acid in it. Another way to find out pH is to know where the water is coming from. If the water comes from a swamp, bog, or other place where the water is pretty stagnant and contains a lot of decaying material, it may be acid. Most water, however, has a pH which is very close to neutral. If the water comes from a river or lake, it is not likely to have a pH that will harm the fish. If the local fish do well in the water, the pond fish probably will do well also.

Liomus Paper. Some farm owners find out their pH by using litmus paper, or pH paper. These are thin strips of paper which have chemicals on them so that they change" color when they are placed into the water. If the water is acid, the paper will turn one color; if the water is basic, the paper turns a different color. The color on the paper is compared to a color chart which will give the pH for that color. There are also electronic meters which measure pH, but these are expensive and not necessary in a field situation.

Hardness. Hardness is the measure of total soluble salts that are dissolved in the water. These salts, usually calcium

(Ca^{++}) and magnesium (Mg^{++}), help the fish grow healthy bones and teeth. Also, the foods the fish eat, like the phytoplankton, need calcium and magnesium for growth. Water that contains many salts is called "hard" water; water that contains few salts is called "soft" water.

Hardness is related to the pH of the water, but unlike the pH, hardness stays constant throughout the day. Hardness can be measured in a laboratory or by using a field kit with chemicals. Hardness should be between 50 and 300ppm in the pond for best fish growth.

There are several ways a farmer can tell if he has very hard water without using chemicals. One method is to look closely at the pond walls where the water line is. If there is a white line on the wall of the pond where the water was touching the pond before the water level fell, there are salts present in the water which have dried on the pond walls. This water probably has a lot of salts. Hardness is important to fish.

Another way a farmer can tell if the water is hard is to wash his hands with it at the side of the pond. If the soap takes a long time to lather, and if the lather does not stay very long, the water is hard. If the water is soft and does not contain many salts, it lathers very easily and is hard to wash off.

If the water is too soft, the farmer can increase the hardness by adding lime to the water.

Alkalinity. Alkalinity is a measure of the acid-combining capacity of the water; or it is also called its buffering ability. Alkalinity measures the amounts of carbonates and bicarbonates in the water. These are materials which mix with acid in the water. The result of the mixing is that the acid is not as strong. Waters which have an alkalinity of 50 to 200 ppm are the most productive for fish. Alkalinity, like pH and hardness, can be corrected and controlled by adding lime to the pond. The relationship among alkalinity, hardness, and pH can be summarized like this:

Low Alkalinity = Low pH = Low Hardness

REMEMBER: THESE THREE FACTORS ARE NOT THE SAME THING, BUT THEY ARE RELATED. IN FISH

PONDS, ALL THREE CAN BE CONTROLLED BY ADDING LIME TO THE WATER.

TURBIDITY

Turbidity is the term for the suspended dirt and other particles in water. Turbidity can be a problem, especially in shallow ponds, if the dirt and particles prevent sunlight from reaching the plankton, so that the phytoplankton cannot produce oxygen. An operating pond can be turbid if there are bottom feeders such as common carp stirring up the bottom mud. Or, turbidity can result from a water source which has a lot of silt in it.

Turbidity can be measured by just looking at the pond water. Or turbidity can be measured by using a device called the Secchi disc. The Secchi disc is also used to determine the total productivity of the pond.

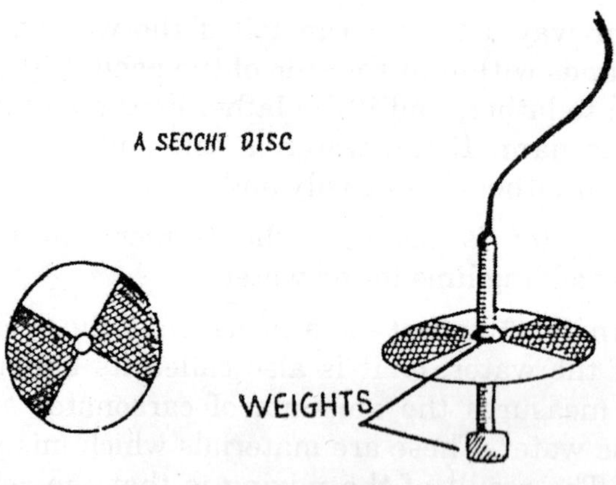

A Secchi disc is about 30cm in diameter, painted white and black or just white, and has weights or heavy objects hanging on it to make it sink straight down into the water. The disc is suspended on a rope or a long piece of wire that is marked off in centimeters from the disc up. A Secchi disc can be made out of wood or metal -- as long as it will sink. The disc does not have to be very complicated. It does not have to be round either. It can be any shape, as long as it has some white paint on It to help it be seen under the water. The disc can be made from a tin can pounded for this purpose.

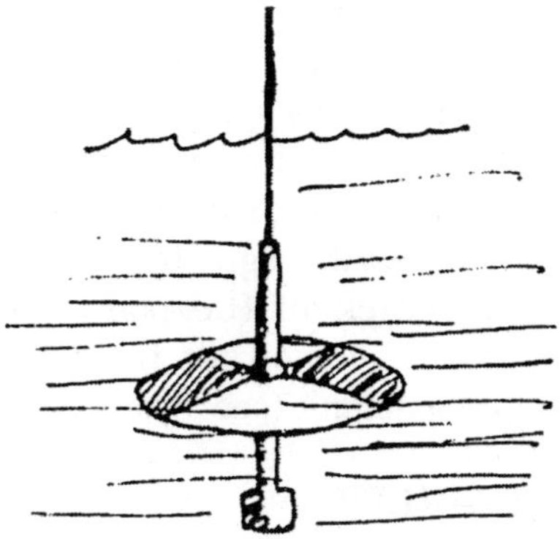

When the Secchi disc goes into the water, it will sink straight down and disappear from sight at some depth. If the disc disappears at 30cm in depth, the pond is not . turbid. If it disappears immediately, either it is very turbid (brown in color), or it is very fertile (productive), if green in color.

Turbidity also can be measured without a disc, but this requires somewhat more experience. The farmer stands in the pond and sticks his arm under the water.

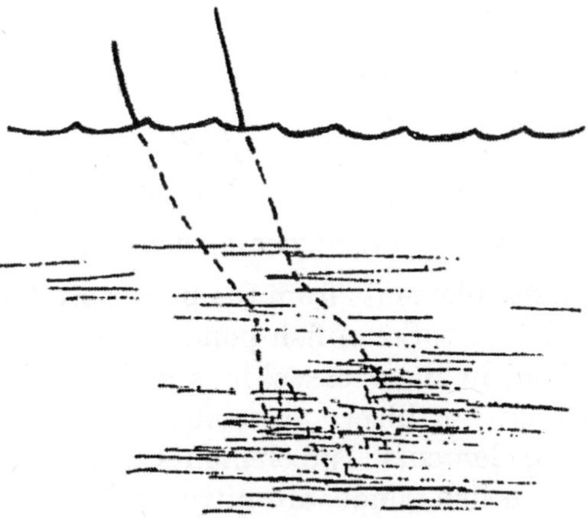

If his hand disappears when the water is about elbow deep, the water is not too turbid. If it disappears the water reaches the elbow, the water is either or very productive. If the entire arm from hand to shoulder can be seen under the water, it is not turbid at all, nor is it very productive (it does not contain enough fish food).

One way to clear up muddy water is to scatter twelve bales of hay per hectare around the edges of the pond. The hay will help to settle the mud and can then be removed easily from the pond edges. However, do not use this method is very hot weather, because the hay will begin to decay very quickly and will begin to use up oxygen in the pond water. If the pond water continues to have a lot of silt in it, the farmer should consider adding a siltation tank (see "Construction").

NUTRIENT AVAILABILITY

All fish require certain elements to grow and reproduce. These essential elements are: carbon, hydrogen, oxygen, nitrogen, phosphorus, potassium, sulfur, calcium, iron, and magnesium. Some other elements called trace elements, are needed only in small amounts. If these elements are missing, or present in too small quantities, the fish will not grow well.

Fish Require a Balanced Diet of Elements

Carbon	*Potassuim*
Hydrogen	*Sulfur*
Oxygen	*Calcium*
Nitrogen	*Iron*
Phosphorus	*Magnesium*

Plus trace elements

Fish get these elements from the pond soil, the pond water, and the food they eat. Some fish ponds lack elements that are necessary to fish. In these cases, it is necessary to add fertilizers to the water. Fertilizers are simply materials which contain the missing elements. The elements most often missing, or in short supply in fish ponds, are nitrogen (N), phosphorus (P), and potassium (K).

Fertilizers containing these missing elements are added to the fish pond to help the growth of the fish and of the plankton, the fish use for food. Fertilization is discussed in the following paragraphs.

Fertilizers

Fertilizers are materials added to the pond to make the water more fertile (productive). As stated before, fertilization is sometimes necessary to help a pond provide the nutrients directly needed for fish and plankton growth. As a major food source of fish, plankton must be kept healthy and in good supply.

Fertilizer supplements the elements the pond gets from its own water and soil. This is especially necessary in ponds made in soil which has used up the nutrients once available.

A WELL-FERTILIZED POND

A pond which has a lot of phytoplankton is often a bright green color. This color indicates a "bloom" of algae. In a normal bloom, the Secchi disc disappears at about 30 cm depth; when the Secchi disc disappears at 20-40cm, the pond is very productive and fertile. No fertilizer is needed in a pond under these conditions. Also, if the farmer places his arm in the pond and his arm disappears from sight at the elbow, the pond does not need fertilizer.

There is one more condition when no fertilizer is needed. Sometimes a pond can become too fertile. If the Secchi disc disappears at only 15cm, the "bloom" is too thick. The thick layer of green blocks the sunlight from the pond and no oxygen can be made by the phytoplankton. In this case, there is too much fertilizer, and the farmer must take off some of the thick layer of algae formed at the top of the pond and stop using fertilizer until the pond has recovered normal fertilizer level.

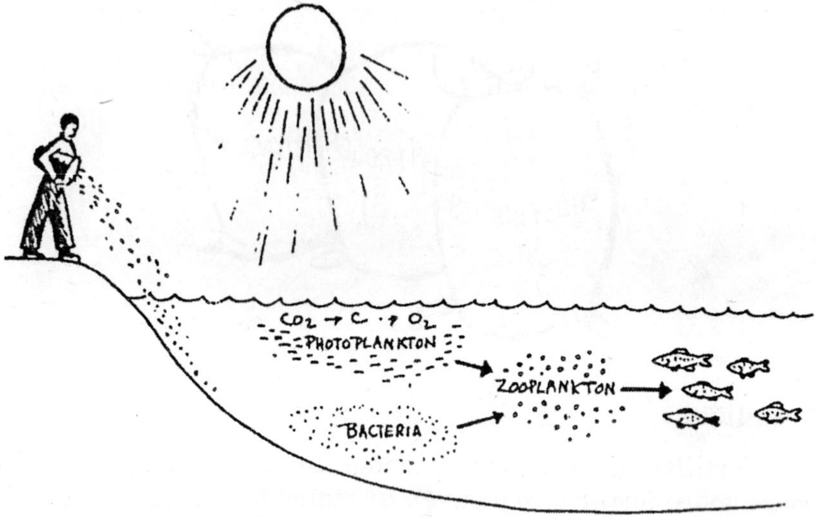

WHEN TO FERTILIZE

If the Secchi disc can still be seen at 43cm, for example, or if the farmer can still see his entire arm from fingers to shoulder under the water, there is not enough plankton. And it is necessary to add fertilizer to the water in order to prepare the pond for the fish.

One other factor which determines the need for fertilizer is the quality of the soil. If the soil is very productive, the need for fertilizer is small; if the soil is not productive, the need is greater. A farmer should know that the fertilizer he uses on his fields, if he uses one, can also be used in his fish pond. The fish pond soil is often very like the soil of the fields around it.

TYPES OF FERTILIZERS

The kinds of fertilizers used in fish ponds vary greatly, depending on the amount of money which can be spent and what is available. Many fish pond owners use organic fertilizers, or fertilizers that come from living things; such as cow dung--because it is available on their farms. Some big pond owners like inorganic fertilizers, or chemicals made by man, like the superphosphates. But these chemical fertilizers are expensive and sometimes hard to get.

Choosing fertilizer can be difficult. The following paragraphs provide more detail about organic and inorganic fertilizers and some guidelines to the proper use of each.

Organic Fertilizers. Organic fertilizers can be plant or animal products, such as:

Vegetable matter. Chopped up manioc, sweet potatoes, or banana leaves kang kong, guinea or napier grass, or other such things that have been allowed to rot for a while. The amounts of vegetable matter used as fertilizer can be as high as 5,000 kg/ha.

Liquid manure. Mostly animal urine containing uric acid, a source of nitrogen. It is washed out of stables into the ponds and used in very small amounts by mixing it with other organic fertilizers such as cow manure.

Household scraps including garbage, grass cuttings, rice husks, and human sewage, also called "night soil".

Animal manure. Includes cow, chicken, duck, sheep, and horse dung. The best way to use this kind of fertilizer is to make a "soup" of it in a tank by mixing it with water. Use the liquid part of the "soup" in the pond. Animal manure can also be placed in a burlap bag hung from a stake in the water. This way, the nutrients from the manure will be released slowly into the water without the manure itself clogging up the pond bottom. If this cannot be done, then pile the manure in the corners of the pond. Do not use too much manure; decaying manure uses up the oxygen in the pond -- particularly in hot, humid climates.

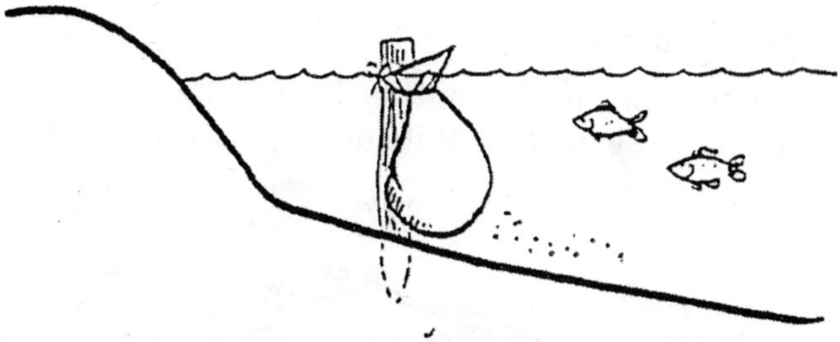

The best way to use these sources of fertilizer is to mix them all together in what is know as a compost pile. A compost pile is simply a pile of these organic materials which has been left to rot. As the materials decay together, they produce a substance which is a very good fertilizer. Compost piles are important: they provide the very best kind of organic fertilizer for fish ponds and, in many cases, they cost nothing.

Making a Compost Pile

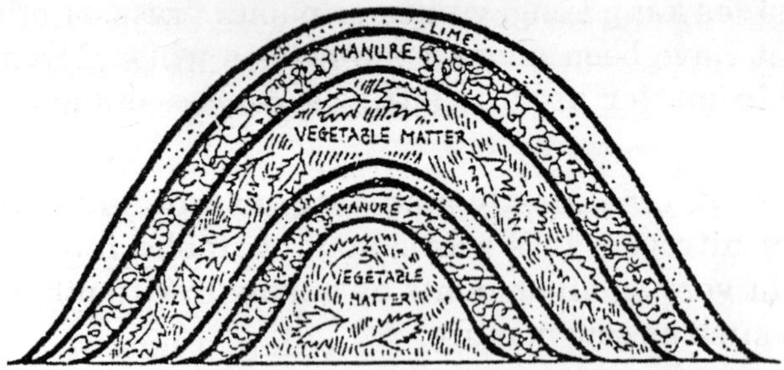

For many years, compost has been made this way:

- Pile organic matter, such as leaves, straw, grass, rice husks or other plant material and household scraps about 30cm high.
- Put a layer of animal manure (chicken, cow, duck, horse, sheep, or goat) on top of the first layer.
- Sprinkle ashes and lime on the manure.
- Repeat these layer of plant material, manure, ashes and lime until the pile is about 1.5 m high and 1.5 m wide.
- Keep the pile moist, but do not let it get wet.
- Turn the pile every three weeks with a shovel for about 3 months.
- Use the pile in 3 months. It will have decayed and shrunk to about 1/10 of its original size.

There is now a faster way to make the compost ready to use as fertilizer.

- Make the same 1.5 m x l.5m pile of plant material, manure, and lime. This time, however, use more household garbage and animal manure. (Animal manure supplies nitrogen, an element used by plants during the decay process. A good compost mixture is about 1 shovelful of manure to 30 shovelsful of the other organic materials).
- Mix the material well. Then cut all of it into small pieces, using a shovel, machete, scythe, etc. The pieces should be about 3 to 5cm long. Cutting the material seeds the rotting process (If animal manure is hard to get, add some inorganic fertilizer containing nitrogen to the compost tile).
- Turn the pile every few days. Use a shovel to keep it well-mixed. Compost piles can get too hot in the middle if they are not turned and mixed. Put a stick into the middle of the pile. Leave the stick in the pile for 3 minutes and then pull it out. If the stick is hot, dry, or smelly, the pile must be turned so that the inside of the pile is now on the outside.
- Keep the pile moist, but not wet. Protect it from the rain. Animal urine can be used to keep the pile moist and helps to add nitrogen to the pile. A compost pile made in this way will be ready for use in only 3 weeks.

When ready, pile the compost in the corners of the pond and restrain it with a screen, or cover the compost with a layer of mud to hold the plant material in place so it does not float into the pond. The compost releases its nutrients into the pond water gradually.

APPLICATION RATES

Fertilizer should be applied at a rate determined by the area of your pond. Area is the length of the pond multiplied by the width. For example, if the pond is 10 m wide by 20m long, it has an area of 200 square meters (mp). This is equivalent to 2/100 of a hectare. The measurements used for pond area are:

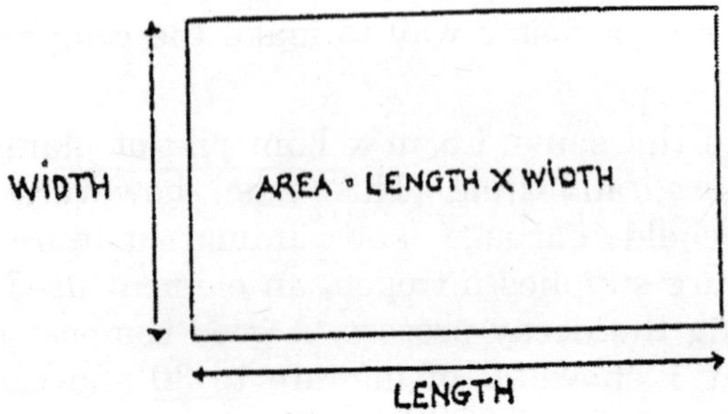

1 are = 100 m^2

1 acre = 40 ares = 4000 m^2

1 hectare = 100 ares = 2.5 acres = 10,000m^2

To fertilize a 200 m^2 fish pond with chicken manure, at a rate of 200 kg/ha, you must only use 4 kg as follows:

$$\frac{200\text{m}^2}{10{,}000\ \text{m}^2} = \frac{\text{x}}{200\ \text{kg/ha}} : \frac{200\,(200)}{10{,}000} = \text{X}\,;\,\text{X} = 4\ \text{kg}$$

Most ponds are not as big as one hectare, so the farmer will have to determine his pond's area before using the manure. It will be hard for most farmers to calculate application rates in this way, but it is probably easy for you to develop some standard measures a farmer can use which are based on the average-sized pond in your area.

Often fish ponds are managed in conjunction with other animals. Stables are built right over the edge of the ponds, and the manure and urine from a certain number of the animals are allowed to fall directly into the pond. This efficient system works well for fish which can use animal manure as fish food. This system gives large crops of healthy fish. Fish ponds which share the area with a number of ducks show high yields of both ducks and fish.

For the first fertilizer added to a new pond, some common rates of application of animal manures are:

Cow dung	1000 kg/ha
Pig dung	568-104 kg/ha
Chicken dung	114-228 kg/ha

REMEMBER: Except for compost fertilizer, only one kind of fertilizer is needed at one time in a pond. Only use one of the application rates each time the pond is fertilized, or the equivalent of one rate. That is, either use 1000 kg/ha of cow dung, or use half cow dung and half chicken dung (500 kg/ha cow + 114 kg/ha chicken dung). After you fertilize keep an eye on the pond. Try not to over fertilize much is just as bad as not enough.

After the first application of fertilizer, application rates do not have to be as high. Many older ponds do not need as much fertilizer because the natural life of a pond tends towards becoming more fertile the older it gets. However, each time the fish are harvested they take part of the pond's productivity with them. That is why older ponds are still fertilized -- even though they may need less fertilizer than new ponds.

Inorganic Fertilizers. Inorganic fertilizers are chemical fertilizers that dissolve in the pond water and provide their nutrients immediately. Originally, inorganic fertilizers supplied nitrogen, phosphorus, and potassium, and they were called the NPK fertilizers. Some typical NPK fertilizers were 8-8-2 (NPK) and 20-20-5 (NPK). This simply referred to the mix of fertilizer that each bag supplied; for example, 8 measures of nitrogen, 8 measures of phosphorus and 2 measures of potassium. Recent studies show that is enough phosphorus is available, the plants in the pond produce their own nitrogen, and that potassium is present already in small amounts in fish. Presently, the only element needed by fish that may be lacking in the fish pond is the element phosphorus.

Now, the most common inorganic fertilizers used in fish ponds are the phosphorus fertilizers -- basic slag, powdered single superphosphate, granular double superphosphate and triple superphosphate. Some of these fertilizers can last as long as three years in the pond, so even though they are expensive initially, they are often used in fish ponds. Research shows that the best fish growth occurs when phosphate fertilizers and organic fertilizers are used together.

Application rates of phosphate fertilizers are:

Basic slag	25-30 kg/ha
Single superphosphate	114 kg/ha
Granular double superphosphate	57 kg/ha

Fertilizers have one purpose -- to provide better growth of fish in ponds. Many organic and inorganic fertilizers are good. Watch the pond carefully for signs concerning a need for fertilizer. As long as the water is a green color, the pond is in good condition. Remember: it is always best to do two things at once--wherever possible use fertilizers which can be used as food by the fish.

Now that the pond has been filled, the quality of the water tested, and the fertilizer added, the last step in preparing for the fish is to make sure that the food supply in the pond is sufficient for the fish that will be put in to the pond.

Foods

It is important to be sure that fish have good food. Feeding and fertilization work together to make the pond successful.

The growth of fish in ponds is directly related to the amount of food available in the pond. The pond must provide all the food and nutrients fish need. But all fish do not need the same kinds of food: different species eat different types of food, and fish eat different foods depending on the stage of their life cycle.

Newly-hatched fry eat from their yolk sacs until the sacs are gone. The fry then eat the smallest phytoplankton in the pond. As the fry get bigger, they can eat bigger foods. Adult fish eat the things that their particular kind of fish enjoy--plankton, higher plants, worms, insect larvae, etc.

TYPES OF FISH FOOD

Fish foods can be natural (those found naturally in the pond) or supplementary (those foods added to the pond).

Natural Foods. These foods are the phytoplankton, zooplankton, detritus snails, worms, insects and insect larvae, small plants like duckweeds and various other weeds and grasses that can be found in a fish pond (See illustrations of Natural Foods at the end of this section). Also, if the fish is carnivorous and eats the flesh of other animals, small fish are also a food source.

Some fish eat all these foods; some prefer only one kind of food. Often a fish will choose one kind of food over another, even though fish if the other food were either of the foods would be eaten by the foods are the best foods for fish. The farmer not available. Natural should encourage, as much as possible, the growth of these natural foods -- through maintaining the quality of his water, proper fertilization of the pond bottom and the water, etc.

Sometimes, however, the farmer must add food to the pond because the pond is not producing enough food for good growth. The best supplementary foods a farmer can put into the pond are extra natural foods. But there are a great number of other foods which fish will eat.

Supplementary Foods. Almost anything can be used as a supplementary food, depending on the fish species in the pond. Typical supplementary foods are: bread crumbs, rice bran, fish meal, ground-up maize, broken rice, soy bean cakes, peanut cakes, corn meal, cottonseed oil cakes, oats, barley, rye, potatoes, coconut cakes, sweet potatoes, guinea grass, napier grass, kang kong, manioc, water hyacinth, wheat, silkworm pupae, and left-over animal feeds and some animal manures.

As stated previously, the kind of extra food depends on the kind of fish. Tilapia, for example, will eat almost anything, including the supplementary foods listed above. This is one reason why they are such very good pond fish. The silver carp, on the other hand, will eat only phytoplankton, even when it is a fish of marketable size. (he farmer rust know what his fish will accept before he puts the fish into the pond.

NOTE DEVELOPMENT WORKERS

Some of these supplemental foods are better at encouraging growth than others. The value of each food is measured in terms of how quickly and well it can help the fish gain weight. The amount of a food that can be converted into fish flesh by the fish is called the conversion ratio. And because these foods are given to help the fish grow, each food has what is known in various places as a growth co-efficient, food quotient, or its nutritive ratio.

The food quotient is figured by dividing the total weight of the food by the total increase in weight gained by the fish over a period of time. This is done as follows:

$$\text{Food Quotient} = \frac{\text{Weight of food given}}{\text{Increase in weight of fish}}$$

For example, a fish weighing 100g is fed a supplementary food at a rate of 5% of his body weight, or 5g per day. The fish weighs 160g at the time of a 30-day period. Therefore, the food quotient of this particular food is:

$$\text{Food Quotient} = \frac{5g(30 \text{ days})}{(160 - 100g)} = \frac{150}{60}$$

$$\text{Food Quotient} = 2.5$$

In other words, the fish has been able to use about 2.5g of food to gain 1.0g of weight a day. This is a good conversion ratio.

The table shown here lists food quotients of some kinds of supplementary foods used with common carp. The lower the value of the quotient, the better the food was used by the fish. For example, dried silkworm pupae help the fish grow faster than do fresh silkworm pupae. REMEMBER : the conversion of foods depends upon the ability of the individual fish to use the food given to it. And that ability differs according to species.

FOOD QUOTIENTS OF COMMON CARP FEEDS

FOOD	FOOD QUOTIENT
Fresh silkworm pupae	5.0 – 5.5
Dried silkworm pupae	1.3 – 2.1
Chironomids	2.3 – 4.4
Fish meal	1.5 – 3.0
Rice bran	5.1
Soy bean cake	2.2
Clam meat	1.3
Cottonseed cake	3.0
Dehydrated blood	1.5 – 1.7
Maize	4.0 – 6.0

Source : Bardach *et al.,* Aquaculture

It will be hard or impossible for many farmers in your area to figure these ratios and quotients. For the farmer who is new to the effort and has few resources, It may be a good idea to direct him to the supplementary foods having the best conversion ratios for his fish.

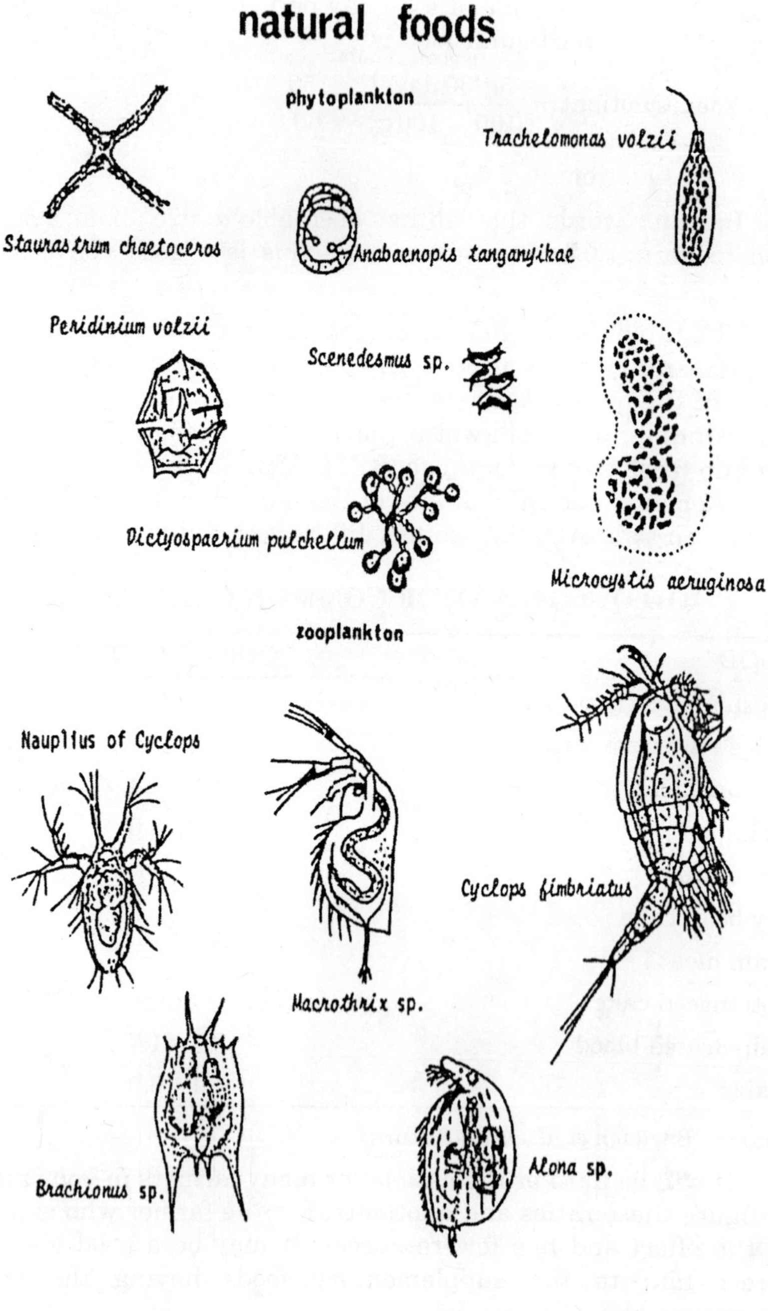
natural foods
phytoplankton
Staurastrum chaetoceros
Anabaenopis tanganyikae
Trachelomonas volzii
Peridinium volzii
Scenedesmus sp.
Dictyospaerium pulchellum
Microcystis aeruginosa
zooplankton
Nauplius of Cyclops
Macrothrix sp.
Cyclops fimbriatus
Brachionus sp.
Alona sp.

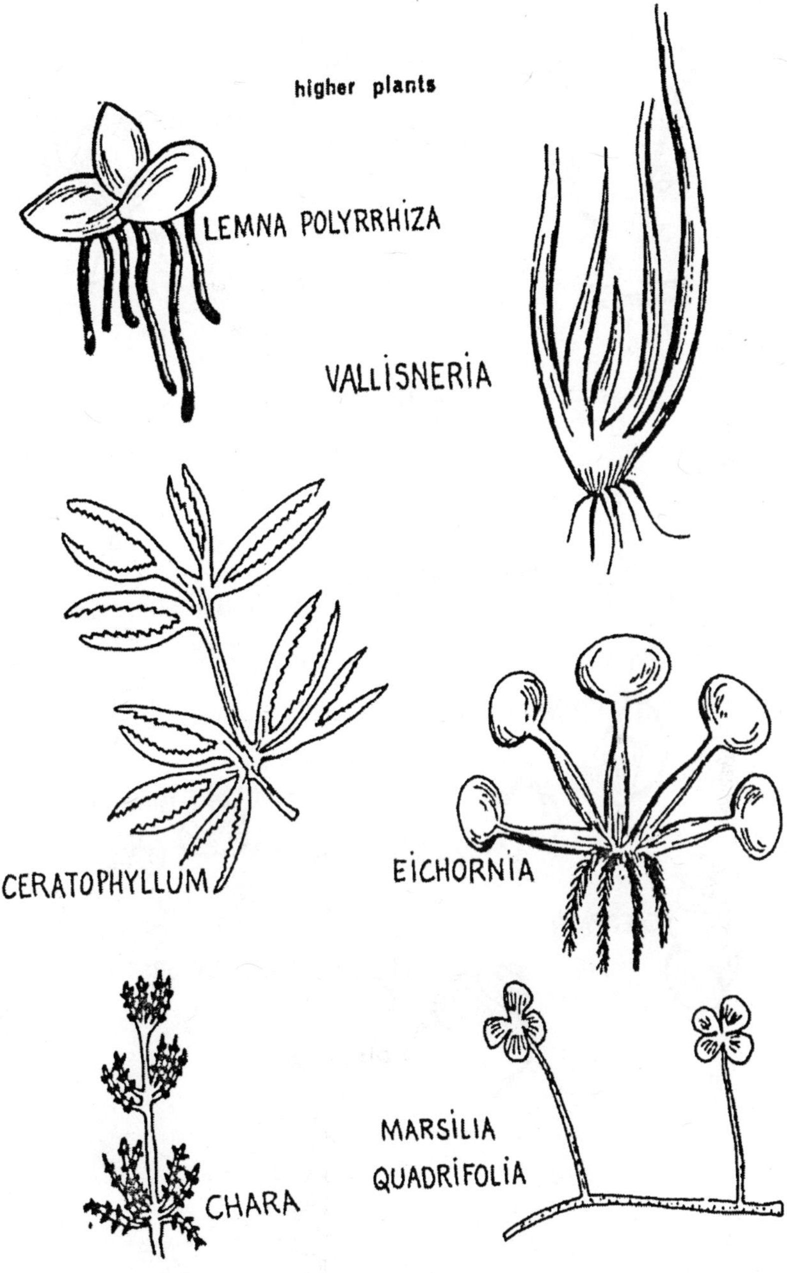
higher plants
LEMNA POLYRRHIZA
VALLISNERIA
CERATOPHYLLUM
EICHORNIA
CHARA
MARSILIA
QUADRIFOLIA

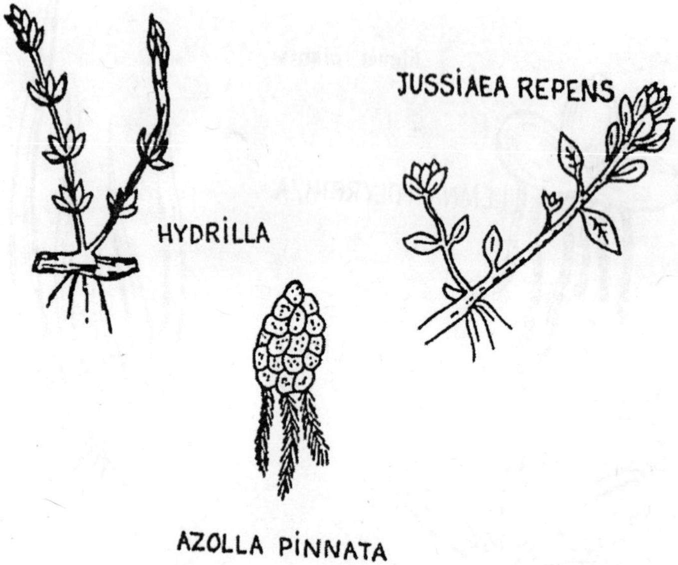
JUSSIAEA REPENS
HYDRILLA
AZOLLA PINNATA

snails

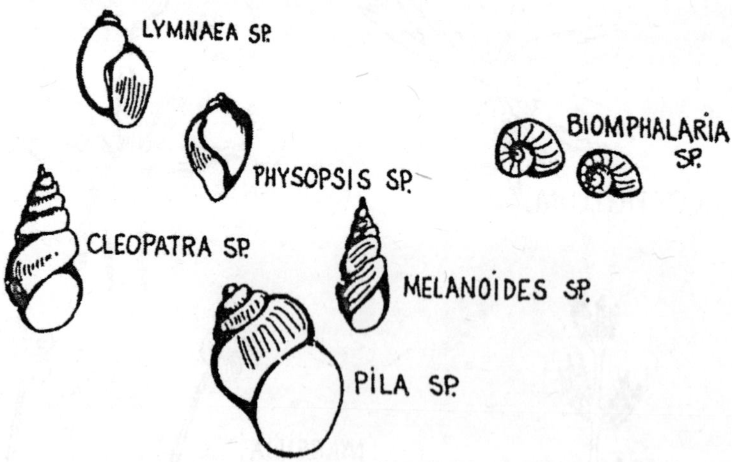
LYMNAEA SP.
BIOMPHALARIA SP.
PHYSOPSIS SP.
CLEOPATRA SP.
MELANOIDES SP.
PILA SP.

6 Managing the Pond

It should be clear by now that much of the success of a fish pond depends upon careful planning. Before the farmer could build the pond, it was necessary for him to think through why he wanted the pond -- for food, profit, or both, what kind of ponds he could build on his land and what kind or kinds of fish are best suited to his climate and pond conditions. Only when all these factors were thought out could the pond be built.

Now, with the pond constructed, fertilized, and otherwise prepared for the fish, the farmer is ready to put the fish into (stock) the ponds and get to the business of raising fish.

Stocking

Stocking is the word used to describe the act of placing the fish (stock) into the pond. The *stocking density* is used here to refer to the total number of fish which can be put into (stocked) in a pond.

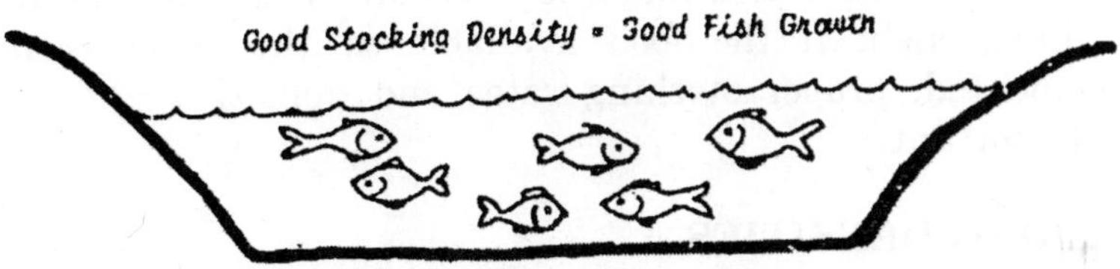

The *stocking rate* is the term used to refer to the number of one species which are put into a pond. Therefore, in a monoculture pond, the stocking rate is the same as the stocking density because there is only one kind of fish.

In a polyculture of Chinese carp, however, the stocking density, or the total number of fingerlings, may be 20,000 per hectare. Of this total, the stocking rate looks like this: grass carp are stocked at a rate of 5,000; 5,000 are bighead carp; 10,000 are silver carp.

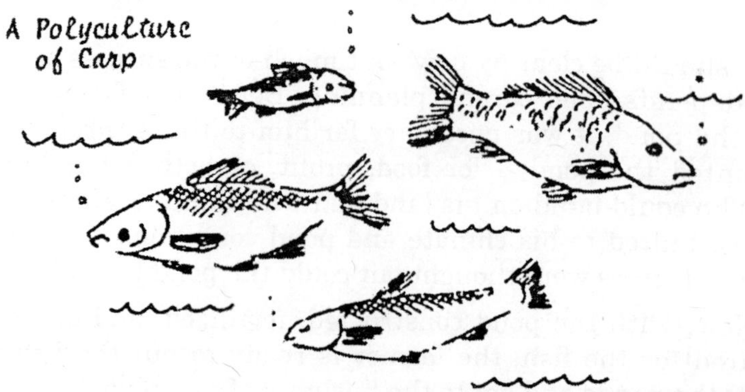

Stocking rate and density are important. There is only enough food and room in a pond for a certain number of fish. The good growth of fish depends upon putting the right number of fish into the pond.

The age of the fish must also be considered when stocking ponds. For example, more fingerlings can be placed in a pond than brood fish, because fingerlings require less food per fish than brood fish. If the food available in the pond is not supplemented, proper stocking rates and densities are even more important.

STOCKING DENSITIES

The farmer must know how many fish he can put into his pond so that he can get the right number--either from the market or from a local stream or lake. He should remember, when he decides upon this number, that some of the fish will die--both when they are put into the pond and later. The following paragraphs provide some guidelines to use when stocking a pond with some of the more common pond fish.

Common Carp. Stocking densities differ with the age and size of the fish. In general, the more volume of water a carp has, the better is its growth. This assumes that the pond contains enough food, and the water temperature is right. The best growth of common carp has been shown with stocking densities of about 10,000 to 20,000 fish per hectare; more with fry; less with post-fingerlings. Some ponds use running water, and in these ponds, they have been able to stock up to 850,000 fry per hectare with only a 20% mortality rate.

Tilapia. Tilapia have been stocked in amounts ranging from 1000 fish per hectare to about 50,000 fish per hectare when supplementary food was provided. But stocking densities really depend on the rates of reproduction of tilapia, and whether they can be separated by sex or not.

Chinese Carp. In general, the stocking rates can only be found by trial and error, and often will be different from time to time, depending upon the availability of fry. In Malaysia, a ratio of carp stocking has been suggested of 2:1:1:3 for grass carp, bighead, silver carp and common carp. This means that if there were a stocking density of 7 Chinese carp, 2 fish would be grass carp, 3 would be common carp and there would be only one each of bighead and silver carp. This is a good stocking rate for this density. The density for a given pond has to be figured in terms of what the pond can support.

Indian Carp. Stocking densities of Indian carp are not widely known. Some densities range from 4,000 to 11,000 fry or fingerlings per hectare, but again, the density depends upon the amount of food available to the fish.

When stocking ponds to produce market-size fish, remember that the more fish stocked, the more food must be available for the best possible growth in ponds.

The following paragraphs describe the proper methods for carrying new stock from the market or river to the pond, and for placing them into the pond.

STOCKING FISH IN PONDS

There are some general rules which apply when bringing fish from one place to another:

- do not handle the fish too much
- make sure the fish get enough oxygen
- keep the fish from getting too warm or too cold
- stock or transfer fish in the early morning when temperatures are lower and the fish are less active.

If fish are stocked so that there is enough oxygen, no temperature difference between the stocking water and the pond water, and they are not touched, the Fish will not be stressed and will survive the stocking. Here are more details concerning the stocking of fish at different stages in the life cycle.

When fry are being moved for a short distance only, for example, from a nursery pond to a rearing pond, they usually are carried in small plastic or metal tubs, or in baskets.

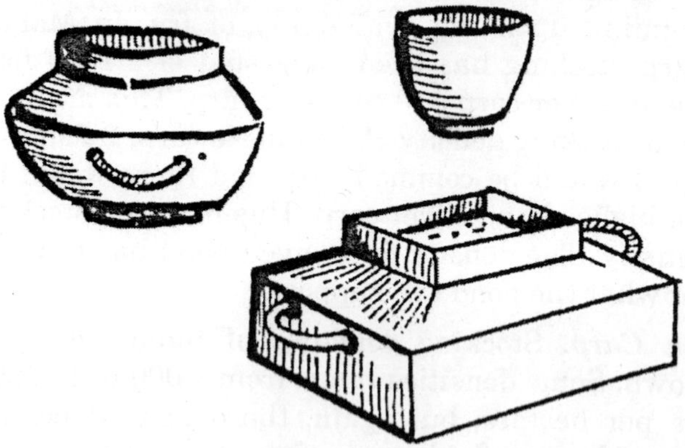

To move fry successfully:

Scoop the fry out of the river or pond in jars, cups, or small nets.

- Put the fry into a bucket of water.
- Carry the bucket to the pond where the fry will be placed.

- Check the temperature of the water in the bucket; it should be the same temperature as the water in the pond where the fry will be stocked.
- Add water from the pond to the bucket slowly -- until the temperature of the water in the bucket is the same as the temperature of the water in the pond.
- Tip the bucket slowly into the pond, and let the fry swim out into the pond themselves.

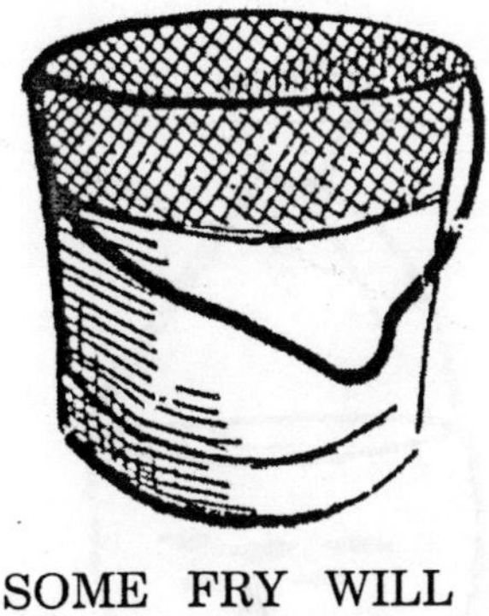

REMEMBER: SOME FRY WILL DIE EVEN WHEN HANDLED VERY CAREFULLY. THIS IS TO BE EXPECTED.

Moving Fry for Longer Distances. If the fry are to be taken from a market or river which requires a few hours travel or a long distance, they must be protected better. One method, which can also be used for fingerlings (and some small adult fish), is to:

- Place fry into plastic bags filled 1/3 with water.
- Fill the rest of the bag with oxygen. The oxygen is put into the bag with a hose placed directly into the water so that the oxygen bubbles into the water.

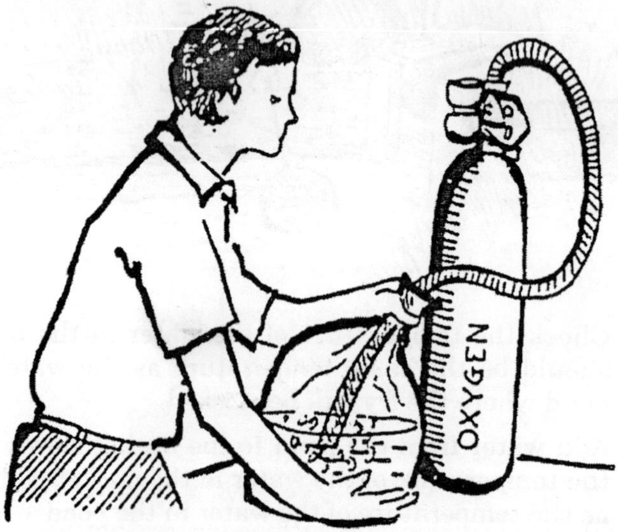

- Tie the bag tightly so that the oxygen does not leak out.

- Place the plastic bags into tin boxes or cardboard boxes or in woven grass bags. These containers give added protection.

- Change the water in the bags after 6 hours. The oxygen will last only that long.
- Make sure the bags do not get too hot and that the temperature of the water in the bags stays at about the same temperature as the water from which the fingerlings or fry were taken.
- Place the bags in the pond unopened until the water temperature inside the bags is about the same as the temperature in the pond.

- Open the bags and let some pond water in.
- Let the bag fill up slowly, and the fish will swim out into the pond by themselves.

This process may take a 1lttle while, but it is far better to take the time than it is to lose the fry. NEVER POUR FRY INTO A POND. This will shock them and kill them all.

Stocking Fingerlings. Fingerlings are stocked in the same way as fry. Always remember that the water in their container must be at the same temperature as the water in the pond. Then let the fingerlings swim out of the container into the pond by themselves. DO NOT POUR FINGERLINGS INTO THE POND. They may die because of the shock of hitting the water or the sudden change of temperature. Some fingerlings will die during stocking. But usually these are the weaker fish. Careful handling will mean less loss of fingerlings, as well as fry.

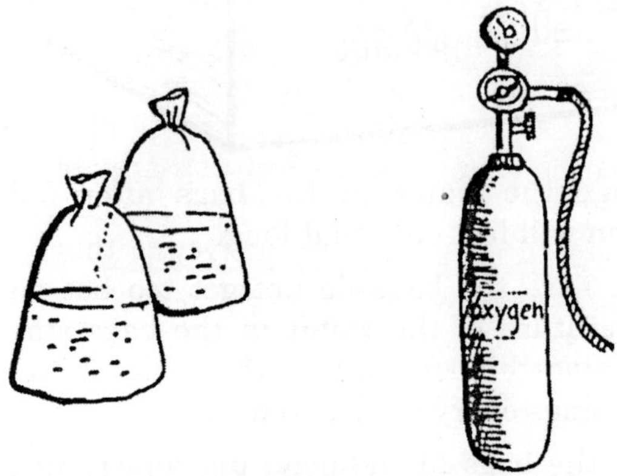

Stocking Adult Fish. Adult fish are a little more difficult to stock than fry or fingerlings. First, they are large (from 0.5 kg up to 3.0kg) and can injure people and themselves by jumping out of containers or ponds when they are being carried or caught. For example, Chinese carp often hurt themselves this way. This problem is controlled by placing a net of some kind on top of the container so they cannot jump out.

To move fish from one pond to another, or from a pond to a container, make a carrying cradle. Use fishnet and pieces of wood or bamboo for handles. The cradle is placed around the brood fish in the water. Then the fish can be lifted out of the water and carried to the new pond or to the container for transporting. There the cradle is released and the brood fish swims away. Brood fish must never be thrown into a pond.

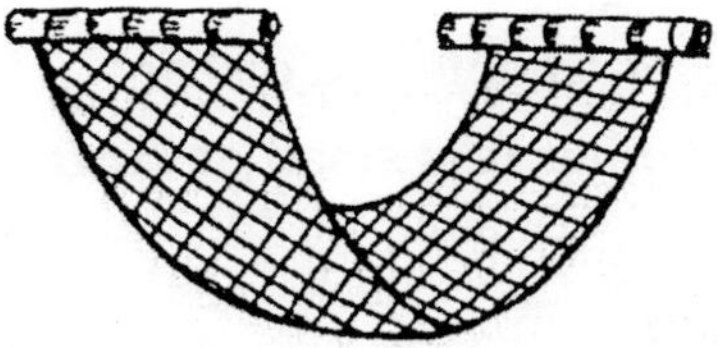

Adult fish often are nervous when being taken from one place to another. Some pond owners even put a hand or a handkerchief over the fishes' eyes when they are carried. Care is necessary when handling, however: brood fish particularly are sensitive to being handled. They bruise easily if they are held tightly, and the bruises can become sites for infection.

Brood fish are often carried in tubs or drums half-filled with clean, well- oxygenated water when they must be carried a long distance. Change the water often and check the water temperature each time. If the brood fish are very active, mix a solution of 1 to 4 grams per liters of urethane in the water. This will make the fish slow and less active, so they can be moved without injury.

Routine Pond Management

After the ponds are stocked, ongoing management of the pond includes:

- feeding and fertilizing as necessary

- keeping the pond in good condition
- watching for trouble and disease

Each pond, whether it is small or large, one pond or one of several, requires supervision in the above areas. And good management requires that checks of the condition of the fish and the pond be a regular part of the pond owner's day. Guidelines for both daily and monthly general maintenance are given here. Then, since fish in ponds are treated somewhat differently depending upon their species, and their stage in the life cycle, more detail on managing fry and fingerlings and managing brood stock is given.

DAILY MANAGEMENT

Ponds and the fish in them must be taken care of every day. It is a good idea to have the pond owner follow a checklist of things to do. Daily care will greatly lessen the chance that something will go wrong in the pond.

A good checklist might look like this:

- check the pond for leaks
- clean filters
- watch fish behavior near the feeding area
- feed the fish
- add fertilizer, if necessary
- watch for predators

IMPORTANT: Check the ponds at the same time each day. Early morning is the best time because oxygen levels in the water are lowest then, and the fish are more likely to have trouble at that time of day--if they are going to have trouble at all.

Each step on the checklist involves certain activities and is discussed in more detail here.

Checking for Leaks. Check all walls, gates, inlets, and outlets. It is possible for a plug on a drainage pipe, for example, to work loose, or partly loose, so that water leaks from the pond. Walls made of hard-packed earth can erode (wash away), especially after heavy rains. Little leaks get larger quickly. It is important to be sure the farmer realizes that in a pond only 2m deep, for example, loss of even part of the water can create problems for the fish.

Cleaning Filters. Again, this is very important. Any filters in the pond must be removed and cleaned of silt, leaves, or other materials that have collected in them. A dirty filter at the outlet pipe could slow down the drainage process.

Watch the Fish. A farmer can tell much about his fish by watching them carefully, If they are swimming quickly and easily around the pond, they are well. If they are waiting near the surface, they are likely to be hungry. If they are gasping for breath at the surface of the water, there is not enough oxygen and the farmer will know he has to act quickly to aerate the water in the pond.

Feed the Fish. Remember : in some ponds it is not necessary to feed the fish extra food. The pond can be made rich enough to fill all the food needs of the fish. However, some ponds and some fish require supplemental feeding. And, sometimes, even a pond which has provided enough food before has to have food added to it.

Supplementary foods are given by:

- spreading the food over the water's surface, as with bread crumbs and rice bran

- placing food inside a floating bamboo or rope feeding ring (which is attached to the bottom of the pond)
- pressing food into dry pellets which float in the feeding ring or fall to the bottom under the ring

Here are some good guidelines for feeding fish which might prove useful to the farmer:

- Always feed the fish at the same time and in the same part of the pond. The fish will learn where to go to get food. Then, when the fish come near the surface of the water, inside the feeding ring, for example, the farmer can see how well they are eating and growing.
- Do not overfeed. Give only the amount of food the fish will take at one feeding. Too much food will not get eaten, but will decay and, therefore, will use up valuable oxygen from the pond during the decaying process.

The amount of food can be found by experience. And of course, the younger the fish, the less food they will need. A farmer is wise to start with a smaller amount of a food. Then, if the fish seem to be waiting near the surface in the feeding area, he will know more food is required.

There are more exact ways to determine how much food to feed the fish. Most pond owners feed fish at the rate of 2 to 5% of body weight per day. Therefore, 100 fingerlings weighing 6g each (a total weight of 600g) would receive 5% of 600g, or 30g of food a day. One hundred fish of breeder size weighing lkg each, (total weight 100kg) would require 5kg of food a day.

Making such measures and calculations is not possible for many farmers. Therefore, it is best that they know which foods to give, how to give them, and how to judge when the fish are or are not getting enough food.

- Feed fish only 6days each week. This will give the fish a chance to feed on whatever food remains in the pond. Too much food can clog the gills of fish, particularly those fish who eat only very fine particles of food.
- Do not feed fish for at least one day before harvesting or breeding them. When the fish eat, they void (empty) the waste from their bodies into the water. This happens even more when the fish are stressed. The combination of food and wastes makes the water turbid and increases the stress that is already placed on fish by the breeding and harvesting processes.
- Feed the right kinds of foods. Some fish will eat almost any of the foods mentioned in the section on "Preparing the Pond." Other fish are not as easy to please. The farmer will have to experiment with supplemental foods. If he gives food one day and it is not eaten, he should stop that food and try another. Again, if he starts with small amounts only, he is not likely to run into trouble. While it is a good idea to test those foods most available to a farmer, here are some guidelines to feeding a number of pond fish.

Common Carp

Common carp feed well on the natural food produced in the pond. However, pond owners often give common carp supplementary food, so the fish will gain weight quickly. Some good supplementary foods for common carp are dried silkworm pupae, fish meal and clam meat. However, these carp will eat almost anything. Supplementary foods such as these are not necessary. The best way to increase common carp growth rates is to fertilize the pond well so that the pond produces a good supply of natural food for the carp to eat.

Tilapia

Not much is known about the feeding habits of some of the tilapia, for example, *Tilapia nilotica. Tilapia mossambica* and *Tilapia zillii* are used to control filamentous algae, which is a habitat for mosquito larvae, thus the tilapia is used to help with malaria control.

Tilapia are hardy and accept many foods. Most tilapia ponds can be managed in much the same way as carp ponds.

Chinese Carp

Chinese carp fry eat plankton, so it is important that they be placed in a well-fertilized pond with a good supply of natural food. Fry can be fed supplementary foods after a while. These foods include egg yolk which is strained through a cloth into the pond, soybean meal, rice bran, and peanut cake. Once the fingerlings get larger, they can be fed like common carp.

Remember, however, that the small pond owner is likely to have Chinese carp as part of a polyculture. If the polyculture has been planned wisely, the Chinese carp will not need to be fed extra food.

Indian Carp

Young fry of Indian carp, like all carps, feed on the plankton in the pond. Normally fish ponds in India are fertilized by draining the pond and drying, then adding a fertilizer made of some animal manure mixed with oil cake at the rate of 200 to 325 kg/ha. This produces a good bloom of plankton for the newly hatched fry. However, it has now been shown that the Indian carp prefer zooplankton, though sometimes they are given supplementary foods. After the fish reach fingerling size, no supplementary food is given.

Note that in any pond, the fish can be kept healthy, well-fed and growing well by making sure the pond is well-fertilized so that it produces its own food. As a general rule, it is better for most small farmers to work at keeping their ponds well fertilized or to find natural foods which can be added to the pond. Most small farmers do not have extra foods to share with fish, but they do have access to organic fertilizer materials, such as manure.

Fertilize the Pond. The section on "Preparing the Pond" discussed kinds of fertilizer, so the farmer should already be familiar with what fertilizers can be used in ponds. Again, the right fertilizer is a matter for experiment and experience.

The farmer has already used fertilizer before filling the pond. Now he must watch the water carefully each day. If the healthy green color of a fertile pond is not there or if the water has become brown, fertilizer is needed. Fertilizers are applied depending upon what kind they are: *REMEMBER:* organic fertilizers do not provide their nutrients right away; inorganic fertilizers work very quickly.

A farmer who uses mainly organic fertilizers would probably be wise to keep some amount of inorganic fertilizer on hand for those times when he needs the fertilizer to work quickly.

Fertilizers are added to the pond in a number of ways;

- Leaves, grass, and animal manure may be left in piles around the inside edges of the pond. This is probably not a good way to fertilize in a hot, humid climate where the faster decay process would result in faster use of oxygen.

- Liquid manures and "soups" are dipped into the pond around the edges or in the deepest water.
- Powdered fertilizers (chicken manure, superphosphates) are broadcast (sprinkled) in a fine layer over the entire surface of the pond.
- Some fertilizers are left on platforms in the pond. The platforms are submerged near the surface of the water and confined behind a screen.

Watch for Predators. Check the pond area for signs of snake holes, rat burrows, eels, and strange fish which may have entered through holes in an inlet screen for example. Any of these can be very dangerous in a fish pond, particularly to a pond containing fry or small fingerlings. Make sure fences which protect ponds from farm animals who might eat grass off the walls or break down the walls of the pond have no breaks in them.

Not each of these things will require much time each day. But a good pond manager will at least check each of these items daily.

MONTHLY MANAGEMENT

Ponds which are managed well day by day will require little other treatment. However, the following things will probably require more careful attention every month or so:

- Check the pond walls. Cut grass which is too long or plant more, if necessary.
- Check the pond bottom. If there is too much buildup of silt and organic matter, shovel or scoop this material out.

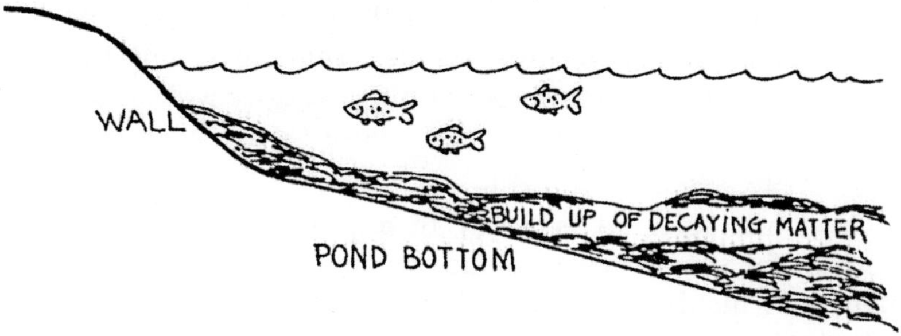

- Check for and remove weeds or other growth which might be a problem at harvest time or when a net is used in the pond.
- Give the walls and inlet and outlet systems an especially careful check for leaks and for blockage. Make sure the water can flow smoothly in and out of the pond, so that if water needs to be put in or taken out quickly, there will be no problem.
- Check the fertility and turbidity of the water. Even a pond well-fertilized at the beginning may need more fertilizer after a month of operation.

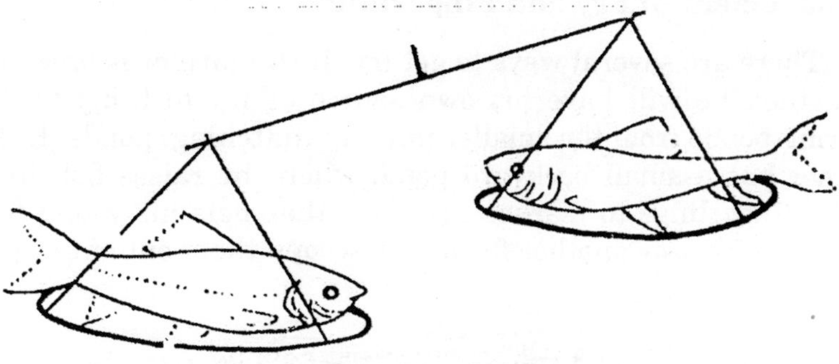

- Check the fish carefully for signs of disease. If all has gone well during the month--the fish have gained weight and their gills are a healthy red color--the chances are that all is well. But the fish should be checked especially carefully for signs of disease each month. (See "Problems of Fish Cultured in Ponds"). It does not take long for a disease to infect an entire pond full of fish.

- Add lime if needed. If the farmer has been adding fertilizer and feeding his fish regularly, but the fish still do not seem to be gaining weight well or moving in the water well, the water quality may need adjusting.

Good management is a key to a good fish harvest. It is important for the farmer to realize this and to work fish management into his daily schedule. But this is not always easy for him to do. In many parts of the world, farmers let their animals manage themselves, i.e., find their own food, etc. This will not usually work with a fish pond. Fish cannot be put into ponds, left alone, and expected to grow and provide food and income. Successful fish pond operation requires active attention by the farmer.

The management guidelines just described apply to all fish ponds, regardless of type of fish or stage of growth. There are, however, some differences between managing a fry or fingerling pond and managing brood stock. So these differences should be looked at more closely here.

Management of Fry and Fingerlings

There are several ways to get fry. If the farmer is breeding fish, then he will have his own source of fry to bring to the rearing ponds from the smaller nursery (hatching) ponds. If the farmer has a small backyard pond, where he raises fish from fry or fingerlings to market size, he either gets his young fish from a market or another farmer or scoops them out of natural waters.

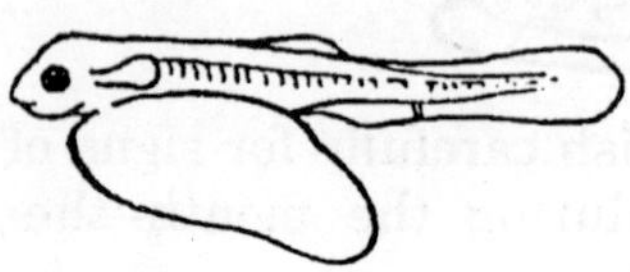

Wherever the young fish come from, it is important for a pond owner to know how many fry or fingerlings he is putting into his pond. If the owner knows how many fish are going into the pond he will know at harvest time how many fish died (the mortality rate) before they were ready for harvest. This information can help the farmer make decisions about his management of the pond. If, for example, more than half of the fish in a pond died between the time they were put in as fry and the time of harvesting for market, too many fish are dying; the farmer ought to find the reasons why before he begins again.

COUNTING FRY

Fry are very delicate and must be handled gently. Here is one way of counting them:

- Take a basin or tub of which you know the size (50-100 1iters)
- Put all the fry into this basin.
- Scoop up fry into a 200-250m measuring cup.

- Count the fry in the measuring cup by slowly and gently pouring the fish back into the basin. Estimate the total number of fry in the basin by setting up a ratio like this.

$$\frac{\text{number of fry in measuring cup}}{\text{number of fry in basin (total)}} = \frac{\text{volume of measuring cup}}{\text{volume of basin}}$$

For example, a measuring cup of 250ml holds 100 fry. Therefore, it is estimated, using this formula, that a 50 l basin full of fry hold 20,000 fry.

Here is another way of counting fry which is somewhat easier because it does not depend upon cups and basins of any particular size.

Put all the fry into an old container--an old metal garbage can, an oil drum, a washtub.

- Get an old milk can, or some other smaller container, and make sure one end is cut off.
- Fill the smaller container with strained water.
- Mark a line on the garbage can to show the level of water being put in.
- Fill the milk can and pour the water into the larger can.
- Continue to fill the smaller can and dump water from it into the larger can.
- Count how many small cans of water it took to fill the larger can as high as the line drawn on the can.
- Fill the smaller can with fry and count them carefully.
- Estimate the number of fry by multiplying the number of fry in the milk can by the number of cans it takes to fill the large container to the line marked on it.

  Therefore, if there were 50 fry in one milk can and it takes 25 cans to fill the larger container to the mark, there are 50 x 25 or 1,250 fry.

Fingerlings are easier to count than fry because they are older and larger. The same kind of measuring system could be set up. But the containers would have to be able to deal with the larger fish. A farmer who has raised his fingerlings from fry should count the fingerlings as he sells them or moves them from a nursery pond to a rearing pond. Then he will know how many survived. If a farmer started with 20,000 fry and had 15,000 fingerlings, 5,000 fry died. But this is a death rate of only 25%--which is not a terribly high figure. Again, the farmer must accept that some of his fish are going to die.

A pond owner who raises fish is more likely to be able to handle fry successfully. Fry are very delicate and must be protected carefully from predators and sudden temperature and oxygen changes. The fry hatch from their eggs in 12 to 72 hours depending upon the temperature and the type of fish. The fry then live off the yolk sac which is attached to them. This sac lasts several days. But then the farmer must be sure that the water provides enough food for the fry.

Many pond owners feed the fry with the yolk of a hard-boiled chicken egg that has been strained through a cloth with water. After a few days of this, the fry can begin to eat the phytoplankton and the zooplankton in the pond. Make sure that there is always enough food for the fry to eat before you transfer the fry to the rearing pond.

For a farmer who has only one new pond, it is probably a better idea for him to start with young fingerlings. This will give more chance of success than starting with fry.

This is not to say that a farmer who has only one pond cannot start his fish crop from eggs or fry. He can. One way this can be done is to keep the eggs in a washtub or large container rather than a pond. The eggs must have plenty of oxygen, so the water must be changed often. Any unfertilized eggs must be removed so that they do not cause infections in the fertilized eggs. Unfertilized eggs are white; fertilized eggs are yellowish red.

Keeping fry in a smaller container is a good idea because it allows the farmer to better control the surroundings. Fry often get bacterial and fungal infections and are a favorite target of birds. Again, the water must be kept rich in oxygen and food which can be eaten by fry.

The care of eggs and fry is very difficult and very important. A farmer who wishes to breed fish must certainly work to gain experience handling delicate eggs and fry. A farmer who wants only a food source in his backyard may wish to take the easier road and start with fingerlings.

The size of fingerlings depends upon climate, water temperature, food given, and the number of fish stocked in the pond. The following are some average sizes and weights common in the Philippines:

	Average Lengths	Average Weights
Milkfish	6.57 cm	2.9 grams
Tilapia	6.33 cm	5.8 grams
	5.64 cm	5.6 grams
Silver Carp	7.39 cm	7.1 grams
Common Carp	7.39 cm	7.1 grams

Fingerlings may be fed supplementary food if it is necessary. Remember that fish usually receive supplementary food which is about 5% of their body weight per day. This was discussed in more detail in the section on preparing the pond, so there is no need to go into detail here.

It probably is a good idea, however, to note again that farmers should proceed slowly when giving supplementary foods. Add only small amounts of food and watch the fish carefully to see how they accept it. And the most important thing is to make sure the pond is producing enough of its own food.

Giving Supplementary Food

If the guidelines for management, discussed earlier in this section, are followed, the fingerlings should grow well. When the fish reach a good size (the size preferred in the farmer's area -- some people like smaller, rather than larger fish), they can be harvested and sold.

A well-cared-for fence protects this farmer's pond from unwelcome visitors.

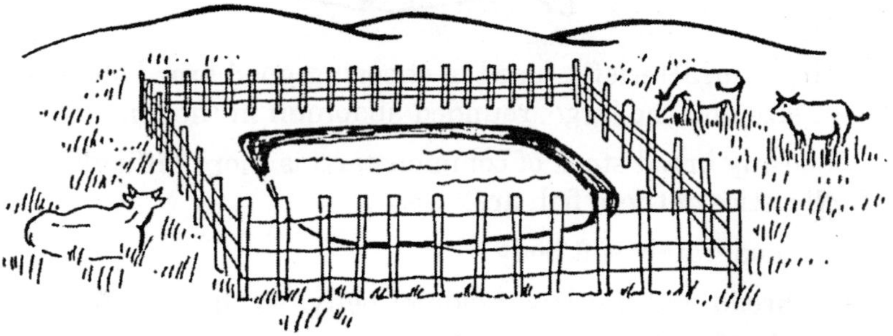

Breeding is the term used to describe the complete reproductive cycle of fish. Successful breeding depends on the health of the brood stock and the ability of the fish to spawn. Spawning describes the actual release of eggs and sperm by the adult fish, and the fertilization of the eggs by the sperm. This section gives information concerning the breeding of pond fish.

Management of Brood Stock

A brood fish is a fish that has reached its full growth and is able to reproduce. The age at which this happens depends upon the kind of fish, the climate, the quality and amount of food. The specific characteristics of brood fish are basically the same for every fish species. In general, good brood fish are:

- well-formed and unbruised
- free of parasites and disease
- lively and active
- a few years old, between 0.5kg and 3.0kg (depending upon species)
- sexually mature (so they can be separated by sex)

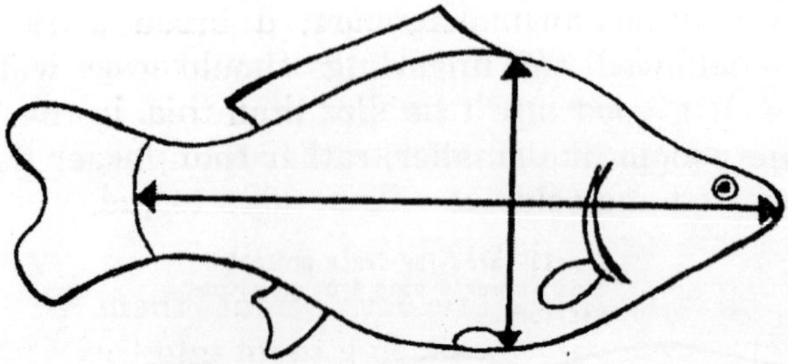

Other characteristics used in choosing good brood stock are relative size and the large, rounded abdomen in the female fish.

Choosing brood stock of common carp is more difficult. The characteristics of these fish are:

- moderately soft body
- broad and flat lower side of belly, so that the fish can stand on its belly
- relatively great body depth compared to length
- broad, but supple, caudal peduncle
- small head and pointed nose
- rather large and regularly inserted scales
- genital opening nearer to the caudal peduncle than in the average carp

In general, the larger the female carp, the more eggs it will produce. A carp of 45-50cm can produce up to 310,000 eggs; a carp of 60-65 cm produces up to 1,507,000 eggs at one time. But older carp (5 year; and up) will have eggs that are not as healthy as those of younger carp (2 years old), so size is not the only factor in choosing good breeders. Good breeders usually are younger fish weighing 1 to 2kg.

Brood fish can be obtained from natural waters by seining (netting) or traps, from fish dealers or fishermen, from other pond owners, or from so that when a government fish farms. Select more males than females, female is ready to spawn, at least one male also will be ready.

The numbers of breeders needed depends upon the size of the brood pond. For example, a carp weighing 1kg needs about

5m² to live and spawn. Therefore, a brood pond of 0.5ha (5,000m²), will hold 1,000 brood fish of an average 1 kg weight. Most brood ponds are much smaller than this, however, so the farmer must calculate the number of fish to place inside. After some experience, the fanner will be able to judge the correct numbers for his pond quite easily.

After choosing the breeders, treat them for possible parasites for disease before placing them into the brood ponds. This treatment is done by placing the fish, one by one, into a bath of 10 ppm of potassium permanganate for 1 hour, them transferring them to a bath of 15ppm of formalin for another 4 to 12 hours. These mixtures can be prepared in washtubs. After the fish are treated, they can be placed into the pond.

Of course, brood fish coming from a source which is known to be uncontaminated and free from disease would not require this treatment. (Further information on treating fish for disease is fauna in "Problems of Fish in Ponds").

The brood stock must be well cared for. If they in good health, the eggs will be healthier. It is probably more important to feed brood stock with supplementary foods than it is to give supplemental food to fish at any other stage of growth. Feed them rice bran, soy bean cakes, or other processed foods at a rate of 5% of body weight per day. They should be managed carefully according to the general guidelines discussed earlier. Remember: brood stock should not be fed for at least one day before they are caught for breeding.

When caught by net, examine the brood stock carefully and handle them as little as possible. Use a cradle to handle and carry the fish from one pond to another. They should be carried to a spawning pond, stocked in the proper manner, and left to spawn. After spawning has occurred, the brood fish should be caught again and carefully carried back and released into their brood pond.

Always remember to treat brood stock well, and never select a fish for spawning which does not show the proper signs of readiness to spawn (See the following information on spawning behavior).

Spawning in fish ponds is done in two ways:

- Natural spawning -- the fish are placed in ponds and left to spawn by themselves
- Induced spawning (artificial propagation) -- methods used by man to make (induce) the fish release their eggs and sperm

Both of these spawning methods have advantages and disadvantages.

Natural Spawning. Fish who spawn naturally require only a well-prepared brood pond. Use a net to seine the pond and choose good breeders. Then introduce them into the spawning pond. Most fish will spawn the first night in the new pond; if they do not spawn, then leave them alone for a few more days. If they still do not spawn, remove them and start again with some other breeders.

Each fish used in pond culture has very definite and very different needs to spawn naturally in ponds. To encourage spawning, ponds can be prepared differently depending upon the fish. Therefore, the best way to prepare is to understand how that fish would spawn in nature. The following describes the natural spawning behavior --in nature and in ponds --of some of the more common pond fish.

THE COMMON CARP -- Spawning in Nature

In China, common carp spawn in the rainy season when the water level and temperature rise at the same time. This rise in temperature and water level is a signal to the carp to begin maturing sexually. When they are fully mature (ripe), they begin their mating behavior, which includes chasing each other in and out of the plants floating on the water surface. The farmer who sees his common carp doing this has a good indicator that his fish are ready to spawn.

When common carp are ready to spawn, the female carp begins to swim in and out of the plants. She then releases her eggs on the plant roots. The male follows her very closely. As she releases her eggs, he releases his sperm (milt); the sperm fertilizes the eggs. Carp eggs are slightly sticky adhesive) and they stick onto the plant roots just under the water surface until they hatch. Depending on the temperature of the water, the eggs hatch in 2 to 6 days.

The new common carp fry feed off of their yolk sacs for another 2 to 6 days, until it is absorbed, and then begin to feed on the zooplankton in the pond water. The carp can spawn all year round in nature, as long as the water temperatures stay high, because a carp is capable of breeding once every two or three months.

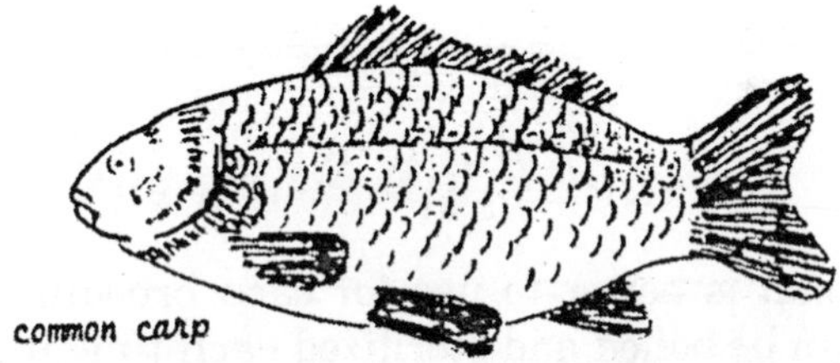

THE COMMON CARP -- Spawning in Ponds

The best way to spawn common carp in fish ponds is to try and reproduce the natural conditions of high water levels and temperature. First the fish are taken from a cool pond and put into a pond with warmer water. Then the water level in the pond is increased. This provides the signal for the carp to mature sexually. When the fish mature, place egg collectors, called kakabans, in the pond, or just some water plants with roots that hang down.

After the introduction of the kakabans, the female fish begin to investigate the fibers. Soon the females will begin spawning behavior and the fish will spawn on the fibers of the kakaban. Because the eggs are sticky, they stick to the kakaban, and the entire kakaban can be lifted and transferred from the breeding pond to the nursery pond.

Important: Common carp are omnivorous; that is, they eat anything including their own fry. It is best to transfer the full kakabans to another pond for hatching.

A kakaban is a floating mat that uses a fiber like inkjuk, or beaten palm bark or leaves that have been shredded into long fibers. These fibers are bunched together and tied in the middle. The bundles are then nailed down between two long pieces of wood or bamboo and floated just under the water surface, with the ends hanging down into the water. This will look like the roots of water plants to the fish.

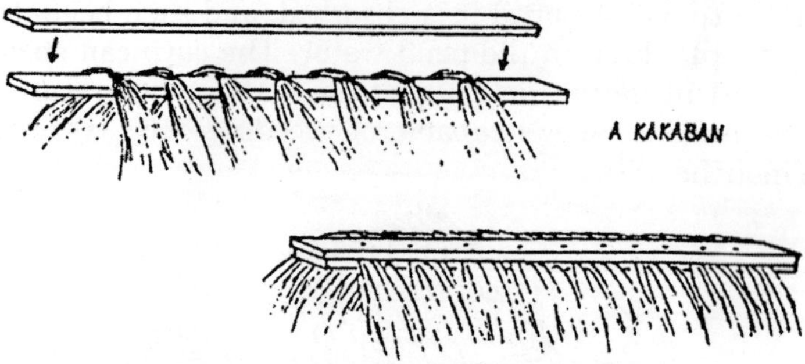

A KAKABAN

A kakaban is better to use for carp breeding than plants because it can be boiled and sterilized each time it is used. This will prevent any fungus or bacteria from attaching the newly-laid eggs.

TILAPIA -- Spawning in Nature

Tilapia spawn every month or so, as long as the water is warm. begins the reproductive behavior by digging holes in the pond bottom or side wall about 35cm across and 6cm deep.

The female will deposit her eggs, about 75 to 200 of them, in the nest, and then the male releases his milt. The female picks up the eggs and the milt in her mouth, so the fertilization of the eggs actually takes place in the females' mouth. Tilapia often are called "mouth breeders."

A drained pond showing tilapia nests.

The eggs remain in the females mouth until they hatch 3 to 5 days. Then the fry stay in the female's mouth until the yolk sac is gone. During this time, the female does not eat.

As the fry grow, they continue to hide in the mother's mouth when they are threatened. The main reason for this mouth-breeding is for protection of the young fish, since the tilapia have relatively few eggs compared to some other pond fish. Tilapia is also a favorite food for a number of predators. Because the fry are so well taken care of by the mother (and even sometimes by the father fish), these young fish are easier to raise than some other species of fry.

tilapia

TILAPIA -- Spawning in Ponds

Tilapia spawn well in ponds. It takes no special equipment or ponds. A tilapia needs only a pond with a loose bottom to spawn. The spawning ponds can be stocked with 25-30 females per 100 m^2 (1/100ha) and about 40-45 males. If the temperature is warm enough, the males will begin digging holes in the pond bottom immediately, and the female will be attracted to the hole and release her eggs. From that point, spawning continues as in nature.

Tilapia will also spawn in ponds that do not have loose bottoms. In these ponds, place large-mouth pottery jars or wooden boxes on sides on their the pond bottom; the tilapia will use these containers as nests.

Young tilapia mature at about 3 months, when they are only 6 to 10 cm long. They can then breed every 3 to 5 weeks, as long as the water is warm. In areas near the equator where the water is always warm, tilapia can breed almost continuously.

When a fish begins to breed, his energy goes into the development of his reproductive organs, not into bodily growth. The main problem with breeding tilapia in fish ponds, therefore, is the rapid reproduction of this fish. Reproduction can be controlled by sorting the tilapia by sex and placing them into separate ponds, or by producing a monosex culture by hybrid crossing. However, these methods can usually be done only by large commercial or government hatcheries where conditions are controlled.

The problem of fast breeding in tilapia ponds can also be controlled by using some natural predators of tilapia in the pond. The predators most often used are catfishes of the genus *Clarias* and, sometimes, eels like *Angulla japonca,* and some other carnivorous fishes like *Serranochromis robustus,* in a polyculture with tilapia that are reproducing. These predators will eat the young fry, allowing the adult fish to continue their growth by having no competition for the available food.

CHINESE CARP -- Spawning in Nature

Chinese carp spawn in the large rivers of China when the spring rains cause the water levels of the river to rise. The eggs are found drifting down the rivers with the current, and they are collected by fry dealers as they drift. The main requirements for hatching Chinese carp eggs are a swift current and plenty of oxygen. Not much is known about their breeding habits in nature, but they are likely to show normal chasing behavior and then spawn, like common carp. Most Chinese carp are cultured by collecting their fry and eggs from the rivers in the spawning season.

CHINESE CARP -- Spawning in Ponds

The Chinese carp are all annual breeders. Good brood stock is chosen in the same way as breeders of common carp. Chinese carp breeders usually are kept in small ponds, separated by sex. When they are sexually mature, it is quite easy to tell them apart, since the males will usually leak milt when handled, and develop other body changes such as serrations (rough edges) on their fins.

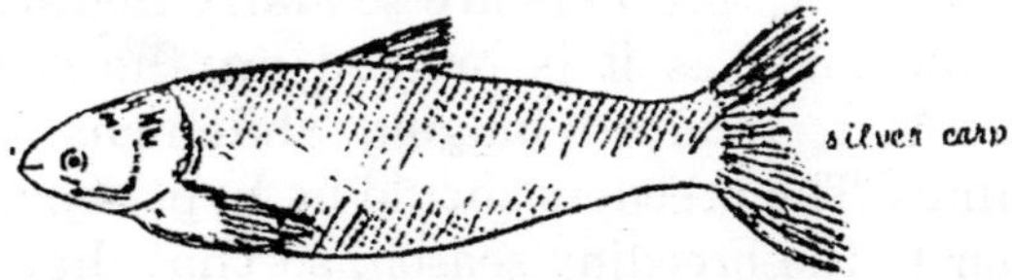

Chinese carp brood stock must be well cared for, like all brood stock. They must be allowed to live undisturbed until time for spawning. However, the Chinese carp must be induced to spawn (see Induced Spawning).

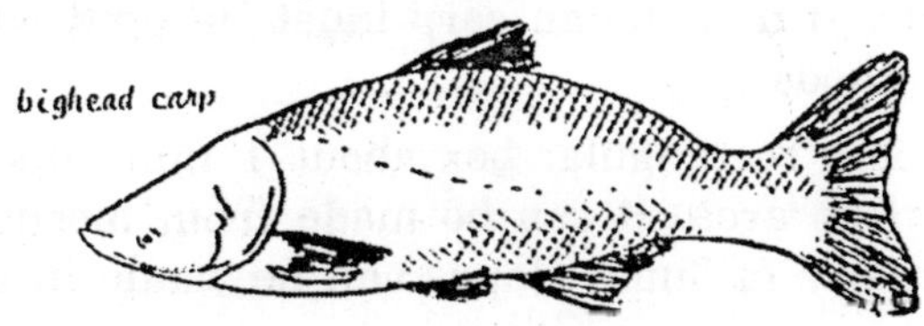

INDIAN CARP --Spawning in Nature

The Indian major carp will not spawn in standing water, so special ponds are built in India to provide a flow of water for these fish. These ponds are built like barrage ponds in upland areas so that the water flows through them. But these ponds are impossible to build in many locations, so the Indian carp often are bred by induced spawning.

In nature, the Indian carp spawn in rivers like the Chinese carp. The eggs are then collected and transferred to hatching ponds.

INDIAN CARP-- Spawning in Ponds

Good Indian carp breeders are sexually mature when milt comes from the male as it is pressed on the stomach. Ripe females have soft, rounded bulging abdomens and reddish genital openings. The breeders should be kept separated by sex in ponds prior to the breeding season, so that they will readily spawn when introduced Into the breeding hapas. Usually one female is placed into a hapa with two males to insure that fertilization occurs. If a farmer can place the breeding hapa into a source of lowing water, he may be able to breed these fish naturally. If not, Indian carp must be bred with Induced spawning methods.

A hapa is a rectangular box about 1 min depth and 1.6 - 6.5m^2 in surface area. It can be made from mosquito netting with a mesh size of 3mm. Hapas can be made in many sizes.

Some other dimensions of hapas used in Indian carp culture are:

91 cm x 91 cm x 183 cm

91 cm x 122 cm x 244 cm

91 cm x 152 cm x 305 cm

91 cm x 183 cm x 366 cm

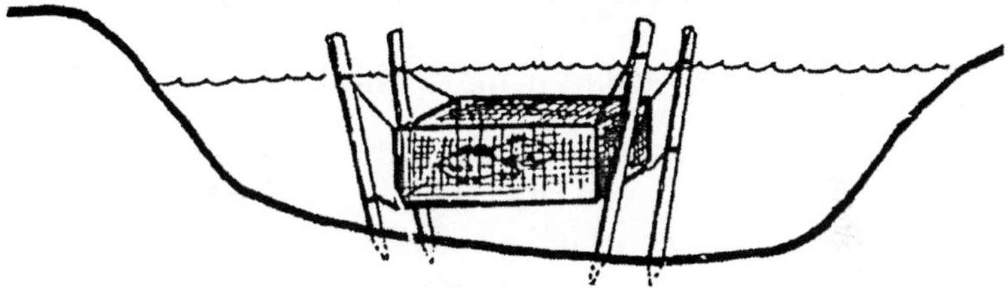

The hapa is held in place inside the pond with stakes of bamboo or other wood. The breeders are put inside the hapa. Kakabans are placed below the water surface, and the top of the hapa is closed so that the breeders do not escape while mating. After spawning, the kakabans can be removed and taken to the nursery pond and the breeders released into the pond. Hapas can be used to spawn other fish as well.

THE GOURAMI -- Spawning in Nature and Ponds

The gourami build nets out of plant materials to lay their eggs. The eggs hatch in about 30 hours. The fry float belly-up for 5 days until feeding begins. The gourami can spawn all year round in warm water conditions.

This is a very good pond fish, and very easy to breed as long as you have a well-fed brood stock. The natural food of the gourami is soft leaves of plants like *Colocasia* and *Carica*. They can also be fed rice bran before breeding. Usually 10 females and 5 males are stocked in ponds as small as 100m? and the eggs float until they hatch.

Breeding in ponds is done by merely placing the breeders together in a pond where there are some marginal plants available for nest building. Once the fry hatch and begin to feed, they can be stocked in nursery ponds.

OTHER GOURAMIS --Spawning in Nature and Ponds

The snakeskin gourami and the three spot gourami build nests made of air bubbles so that their eggs float. The kissing gourami scatters its eggs, which are free-floating.

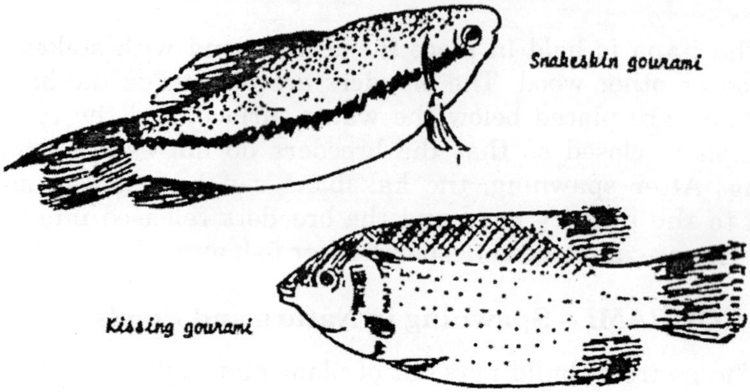

To breed the snakeskin and three spot gourami, place the ripe fish into a well-oxygenated pond that has a good growth of aquatic vegetation, particularly *Hydrilla verticillata*. These fish will continue to spawn as long as the water temperature stays at 26 -28°C. Hatching takes place about 2 days after spawning, and the fry absorb the yolk sac within 3 to 7 days.

The kissing gourami spawn at 6-month intervals and spawn within 18 hours of stocking in the pond. Some of the eggs may be eaten by the parent fish, so there must always be abundant vegetation in the spawning pond to prevent this. The eggs hatch in 2 days and float on the surface for 3 to 4 days. The new fry eat the decaying plants and plankton in the pond.

CLARIAS CATFISH -- Spawning in Nature and Ponds

Clarias macrocephalus spawns during the rainy season in nests on the bottom of natural waterways, while *Clarias batrachus* spawns in horizontal holes in the banks. Hatching takes place after 20 hours at 25-32°C. The fry are then collected by hand net from the nests. There are 2,000 to 15,000 fry in each nest.

Clarias catfish will spawn naturally in ponds, but induced spawning methods may be used if necessary.

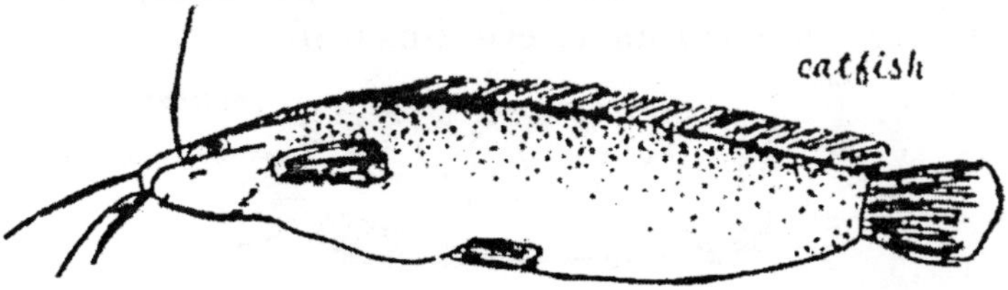

THE TAWES -- Spawning in Nature and Ponds

The tawes spawn in the rainy season. Tawes ponds usually are about 200 to 500m^2 and about 50cm deep. The ponds should be dried for 5 days before they are filled, and the spawners should be introduced when the pond is half full. Tawes need well-oxygenated water that has a strong current to spawn. Mating occurs at night; then the current should be turned off and the eggs spread out evenly on the pond bottom. The eggs hatch in two to three days. After 20 days, the fry can withstand the current, and it should be turned on again. Tawes females produce about 20,000 fry each.

HETEROTIS NILOTICUS - Spawning in Nature & Ponds

This species is normally light-colored, but during the breeding season it changes to dark brown. The spawning of *Heterotis niloticus* in nature begins at the end of the dry season when water is very warm. The fish splash in the shallow water among the weeds in the ponds to begin their breeding behavior. Then the male builds a nest of weeds in water that is 10-45cm deep. The nest is made in a depression that is 15cm deep and 60 -100cm wide. The nest has a grass wall at its outer edge which keeps other fish out of the nest. To get in and out of the nest, *Heterotis niloticus* jumps over this wall.

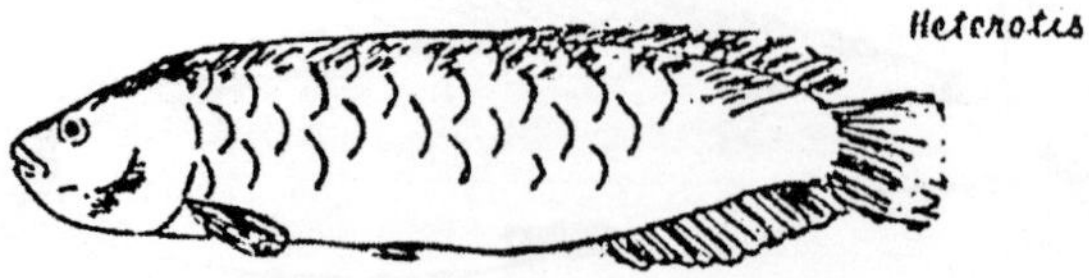

The eggs of *Heterotis* are about 3mm in diameter, and are laid in the bottom of the nest and then fertilized. One of the parent fish is always in the nest to circulate water over the eggs (to give them oxygen). The eggs hatch in 4 -5 days. The fry travel in a "school" and stay with their parent fish for several months after hatching. The fry are very delicate, and should not be handled for a while.

EELS -- Spawning in Nature and Ponds

The eels used in Taiwan *(Anguilla japonica)* spawn in the sea, and the fry (called elvers) swim upstream where they are collected by dealers. Eels, are stocked in rates of up to 25,000 fry/ha along with other fishes, and must be fed supplementary feeds like pellets of trash fish. It is not recommended that a beginner work with eels because they must be fed protein and are not very efficient converters of food.

Eels cannot be bred in ponds.

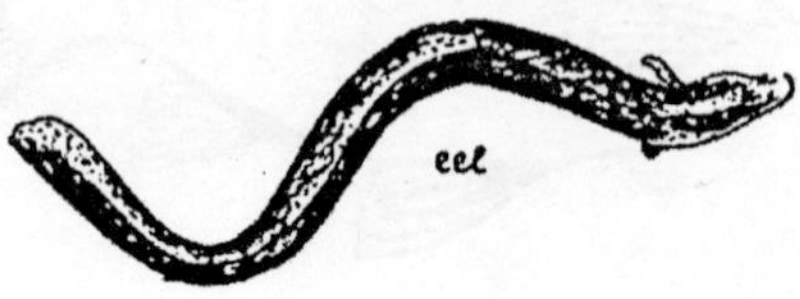

MILKFISH -- Spawning in Nature and Ponds

Milkfish spawn in saltwater during the rainy season. The fry are caught along the shore line at breeding season (which corresponds to the rainy season) and then transferred and acclimatized to freshwater ponds. This is done for the most part in the Philippines and in some other Southeast Asian countries like Indonesia and Taiwan.

Milkfish cannot be bred in ponds.

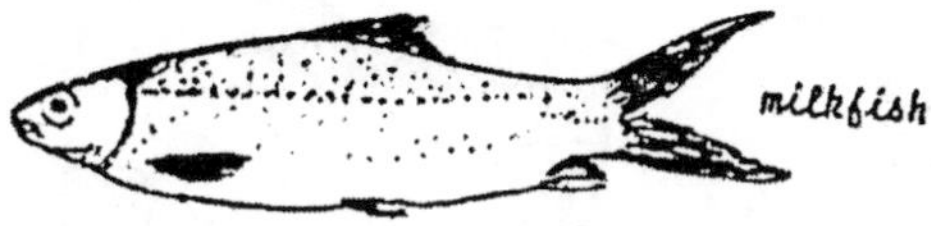

STRIPED MULLET -- Spawning in Nature and Ponds

The striped mullet is a saltwater fish, and spawns in the sea. The fry are collected as they swim upstream.

The mullet can be induced to spawn by hormone injection, but this is very difficult and certainly is not recommended for a small fish pond owner.

Induced Spawing. Induced spawning means making the fish produce eggs and milt when they will not do so naturally. Induced spawning is done when the pond conditions cannot be made to encourage natural spawning, or when the fish are not ready to spawn when the farmer wants them to spawn.

Spawning can be done by three methods:

- hormone injection
- hormone injection with stripping
- stripping

Each of these methods has advantages and disadvantages.

Hormone Injection. Hormone injection is the most common method of induced spawning, and it requires certain kinds of equipment:

- hypodermic needle and syringe
- mortar and pestle
- saline solution or distilled water

- centrifuge
- test tubes
- dissecting kit

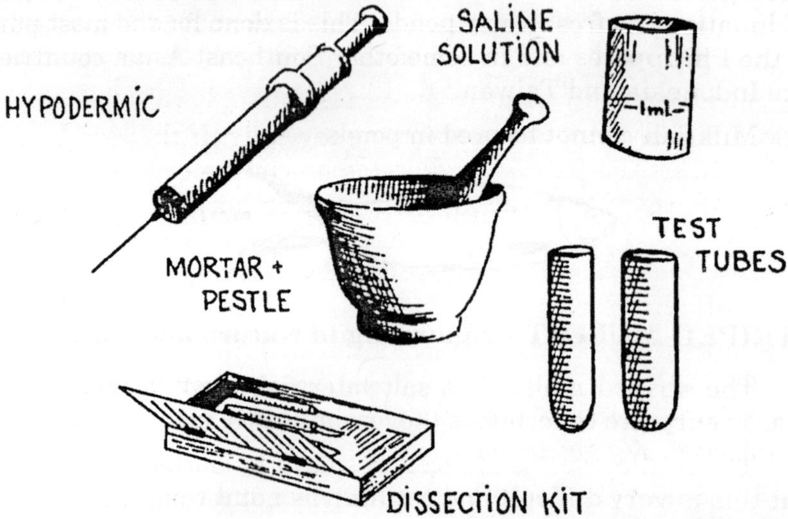

This technique uses the pituitary gland (the hypophysis) of the fish. This gland contains the substances (hormones) that trigger the reproductive organs of the fish to start developing. When these hormones are taken from a ripe fish and injected into a fish that is ripe, but has been unable to spawn, the injected fish will spawn in 6 -12 hours.

The ripe fish must be killed to get the pituitary gland out. This must be done very carefully. The gland is very small: less than 1 mm in diameter in the common carp, which has a relatively large pituitary. The pituitary gland is a round, yellowish-red organ located in the brain pan of the fish. Here is the method commonly used to take the gland from the fish:

- Use a mallet or dull knife.
- Hold the fish near the head with one hand.
- Hit the fish above the eyes at the point where the skull begins. This will kill the fish.
- Make sure to hit straight and up a little. A solid hit should dislodge the skull.

- Slit the skin around the front and sides of the skull, then lift up the top of the skull and fold it back as if it were a hinge. The brain is attached to the top of the skull; by folding it back, the underside of the brain is exposed. The pituitary gland is located in the middle part of the underside of the brain.

If located in this way, the pituitary gland is relatively easy to find. However, this must be done carefully. If the skin is cut too much, or the fish is handled too much, the contents of the brain will move and the pituitary will be hard to locate. The brain contains a number of fat deposits which are yellowish and could easily be confused with a pituitary by someone who was not familiar with that gland.

Most farmers will not be interested in doing hormone injection spawning. But you should be familiar with and be able to do it. Steps for processing the pituitary gland and giving the injection are given below:

- Select the fish you want to spawn and weigh them.
- Select the fish that will be killed for their glands and weigh them. Always match the weights of the donor and recipient fish. If a donor is 1.5 kg and the recipient is 3kg, use pituitaries from two 1.5 kg donors.
- Kill the fish, as outlined above.
- Remove the pituitary from the fish matching weights (or use 2 to 3mg of dried pituitary gland for every kilogram of body weight).
- Place the pituitary gland into the mortar.
- Grind the pituitary with pestle until it is a pulpy mass.
- Wash the pituitary into a test tube with 1 milliliter distilled water or saline solution.
- Place the test tubes into the centrifuge.
- Centrifuge the glands for 5 minutes.
- Remove the test tubes from the centrifuge.

- Draw up the liquid portion from the test tube into the hypodermic needle, leaving the pulp of the gland in the bottom of the test tube.

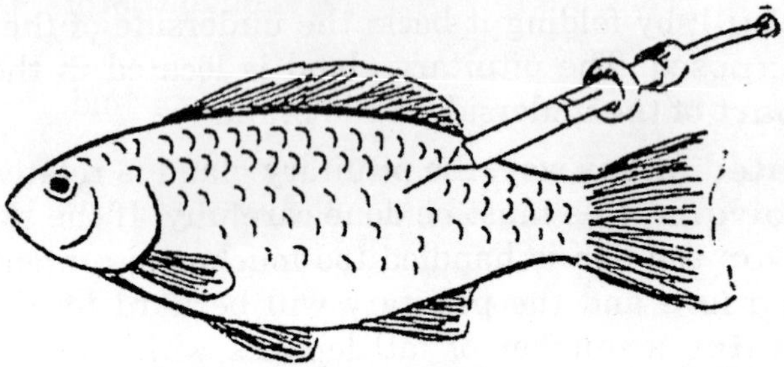

- Inject the fish above the lateral line behind the dorsal fin, just underneath the scale.
- Place breeders into the breeding pond.

All of the materials used in hypophysation can be found or made easily. A simple centrifuge can be constructed very easily from a hand drill (see next page for instructions). If a centrifuge cannot be found or made, the fish may be injected with a whole pituary gland. The gland may be dropped into the syringe, water added, and injected into the fish as outlined above. The force needed to push the gland out through the needle will crush the gland as if it were being ground with the mortar and pestle, and this will allow the hormones to be released.

After the injection, the female fish will begin to develop her eggs until they are ready for fertilization by the male. En some fish, it is necessary to inject the female twice with varying amounts of pituitary extract (see Chinese carp) and the male, once. After the injections, the fish are treated in the same way as in natural spawning.

MAKING A CENTRIFUGE

Tools and Materials:

- 1 hand drill
- 1.20 cm piece small wood (or bamboo) dowel rod

- 2 metal cigar tubes (or plastic, or rubber hose with clamps)
- 2 pieces medium wire (long enough to wind around the tubes 6 or 7 times)
- 1 piece heavy wire about 9cm long (old coat hanger will work)
- cotton balls or pieces of soft material
- 2 test tubes or small clean glass bottles
- string, tapes, and nylon fishing line

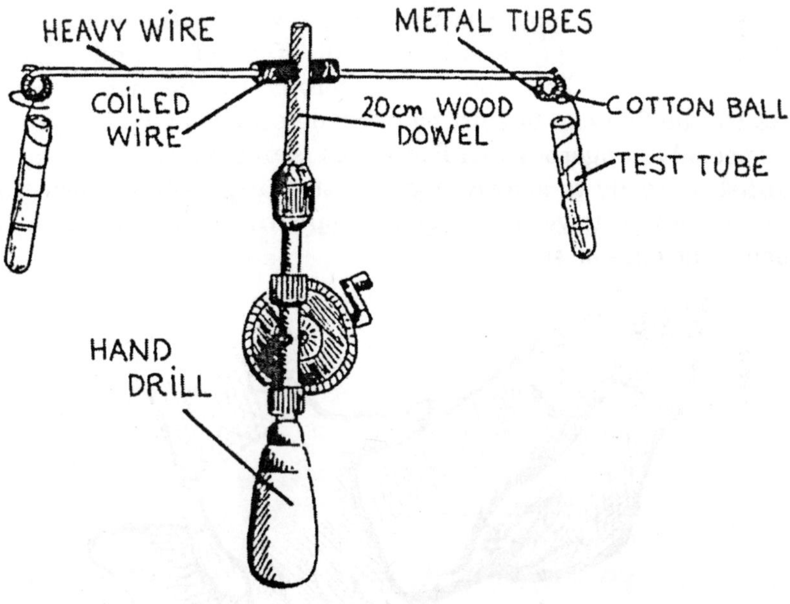

Construction Steps

- Drill hole through one end of dowel rod about 1cm from the end.
- Remove drill bit from drill.
- Insert dowel rod into drill bit hole, leaving end with newly drilled hole at opposite end on top.
- Insert piece of heavy wire through the hole in dowel rod.

- Bend the ends of the wire into loops.
- Secure the wire on either side of the dowel rod with tape to keep the wire from slipping through the dowel rod hole.
- Wind medium wire around each cigar tube leaving about 2.5 cm of wire free at the top of each tube.
- Attach tubes to heavy wire by bending medium wire (left over from step just completed).
- Place a small cotton ball in the bottom of each tube to cushion the test tubes.

Stripping. Stripping is the term given to the method of actually pushing eggs and sperm out of the fish and mixing them in a dish. This can be dangerous to the fish, mostly because the fish can be hurt by pressing on the belly. Stripping is especially dangerous to a fish which is not ready to spawn. If the fish is ready to spawn, a gentle stroking motion down the side of the fish towards the genital opening will be enough to release the eggs or sperm.

First the eggs are stripped into a dry dish. Then the milt is stripped into the same dish. Mix the eggs and milt gently with a feather. Add water to the dish so that fertilization can occur. After a few hours, and a few changes of water in the dish (to provide eggs with oxygen), transfer the fertilized eggs to the kakabans and allow them to hatch as normal.

There are other variations of stripping that are worse than the one outlined above. One method involves killing the female or male, or both, and removing their reproductive organs and then mixing the eggs and sperm by hand. Not only is it necessary to kill both breeders, but if the eggs and sperm are not ripe (mature) and ready for fertilization, no fry will hatch.

Stripping with Injection. Often stripping is done after the fish have been injected with hormone extract. The fish are injected, and the eggs are allowed to develop. Then the fish are stripped into a dry dish, etc. Stripping with injections works fairly well. But of the three methods of induced spawning described here, the best is just to inject the fish and let them spawn by themselves in the pond. The following paragraphs give directions for induced spawning of some important pond fish.

INDUCED SPAWNING OF CHINESE CARP

The Chinese carp must be induced to spawn by hormone injection. The normal dosages for big head, black, mud, and silver carp are 2 to 3 mg of dried pituitary or 3 fresh pituitary glands for every kilogram of female fish. That is, if the female silver carp weighs 2.5 kg, 5 mg of dry pituitary or 6 fresh pituitary glands are needed to ripen her eggs. Or HCG (human chorionic gonadotropin) can be used at dosages of 700 to 1000 IU (international units) per kilogram. But HCG is expensive and certainly not available to everyone. Grass carp need higher dosages (3 to 4 mg dried pituitary per kilogram of body weight). Inject only a fraction (1/10 to 1/4) of the total for the first dose; then, follow it with the rest of the dose, 6-24 hours later.

After injection, put the breeders into the breeding pond. The temperature should be about 23-29°C to encourage spawning, and the oxygen content should be at least 4 ppm. It is best to put in two males for every female. Let the fish spawn on their own; they will spawn within a day. Remove the breeders after spawning.

Hatching Chinese carp is complicated. Chinese carp eggs need a constant supply of clean, well-oxygenated water flowing from the bottom up through the eggs to stimulate the hatching. Some types of hatching bags have been developed for use. One kind of bag hangs from a rack down into the nursery pond or a trough, and water is bubbled up by pipes from the inflow pipe. These bags have an advantage in that once the fry are hatched, they can easily be transferred without touching them at all. This is good, because Chinese carp fry are very sensitive to handling stress.

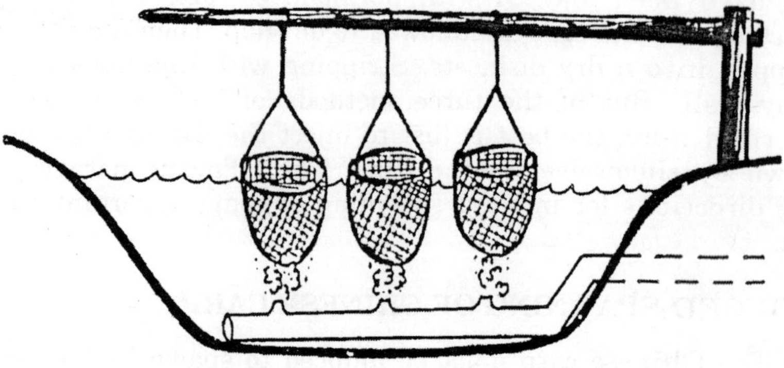

After the carp spawn, the eggs are collected by net or by "raining the breeding pond, and they are placed in the hatching bags (or shallow trays) as soon as they have hardened after fertilization (1 to 2 hours). The eggs hatch in 1-2 days depending on the temperature, and then absorb their holk sacs in another 3-6 days.

As soon as the fry absorb their yolk sacs, they should be transferred in the hatching bags to nursery ponds. The nursery ponds should be 0.5 to 1.0 m in depth and the oxygen level should be at least 4 ppm for good fry growth.

The spawning of Chinese carp is a very complicated business and is usually done inside carp hatcheries so that all conditions can be controlled. In China, the carp hatcheries sell their fry to fish pond owners who then raise them to marketable size. For most farmers, common carp is a much easier fish to work with and is just as valuable for food as are Chinese carp.

INDUCED SPAWNING OF CLARIAS CATFISH

The *Clarias macrocephalus* fishes are injected with pituitary extract at a rate of 13 to 26 mg/kg at 25-32°C. Spawning occurs within 16 hours. Larvae (fry) absorb the yolk sac in 5 days, and are transferred and reared in ponds only 18 cm deep. The best food for fry is zooplankton, but after 2 to 3 weeks, trash fish may be added. They can be fed rice bran as well, and later on a mixture of trash fish, rice bran, and broken rice. In Thailand this sort of production gives yields of 97,000 kg/ha per year. *Clarias* catfish are used in fish ponds throughout Southeast Asia now, and are enjoyed for their good taste.

INDUCED SPAWNING OF INDIAN CARP

If you cannot build a fish pond like a barrage pond or spawn the Indian carp naturally in ponds, they can also be induced to spawn by hormone injection but this is very difficult to do. Induced spawning is dependent on the dosage and the stage of maturity of the breeders. Breeders should be about 2 to 4 years old, and weigh 1.5 to 5.0 kg. Females are injected twice, once with 2 to 3 mg of pituitary gland per kg body weight, and then, after 6 hours, with 5 to 8 mg/kg. Males are injected once, at the time females get their second dose, with a dose that is equal to the first dose given to the females. After the second injection, the fish are placed together in breeding "hapas" and spawning takes place within 3 to 6 hours. The breeders are put inside the hapa, kakabans are placed below the water surface, and the top of the hapa is closed so that the breeders do not escape while mating. After spawning the kakabans can be removed and the breeders released into the pond. The eggs should be transferred to deep hatching hapas where they will hatch in 15 to 18 hours at 27°C. However, this induced breeding does not work as well as Chinese carp breeding, so most Indian carp fry are still caught and collected in natural waters.

INDUCED SPAWNING OF COMMON CARP

Sometimes common carp will not spawn in ponds, and they are injected. The amounts needed for common carp are determined by the fishes' weight. Usually the common carp is injected only once with pituitary extract from a fish that has the same weight as the injected fish. The male is not injected. After the injection, the fish art placed into the breeding pond. Usually a good female breeder will weigh 1 to 2 kg. This one large female is placed with 1 or 2 males, so that the total weight of the males is approximately the weight of the female. If you have a female of 2 kg, you can use two males of 1 kg each. The more males, the greater the chance that fertilization will occur. If you have a large breeding pond, you can place about 5 or 6 large female fish and 10-15 males to insure that all the eggs are fertilized.

Carp will only respond to pituitary injections from other carp. However, many other fish will respond to the pituitary gland of common carp, so often carp are kept just to serve as donors of this gland in other induced spawning attempts. Also, carp glands are relatively large and easy to find, compared to the glands of other fishes, and can be stored for later use by drying, freezing, or powdering. Carp glands can be preserved by placing them in 100% dry acetone, then cooling them by placing that jar they are placed in, into an ice bath. Every 12 hours, the acetone should be changed, for a total of four times. Then the pituitaries are air-dried, and stored in an air-tight container. This is called the alcohol drying method; glands preserved in this way can still be used after 10 years.

7 Harvesting Fish

Harvesting is the collection of fish from a pond for sale at market, or for cooking and preservation for family use. Harvesting can refer to collecting all the fish or to taking out only some of the fish (this happens often in tilapia ponds having both young and adult fish).

If the pond can be drained, harvest the fish by draining the pond into the catch basin and collecting the fish with a scoop net. If the pond cannot be drained, drain out as much water as possible and use a series of nets to catch the fish.

Types of Nets

There are different kinds of nets which can be used in ponds. Some nets, such as the one shown here are gill nets. Gill nets often have mesh sizes from 2-3cm; they are often used to harvest the largest fish in a pond and leave the smaller fish until they grow larger.

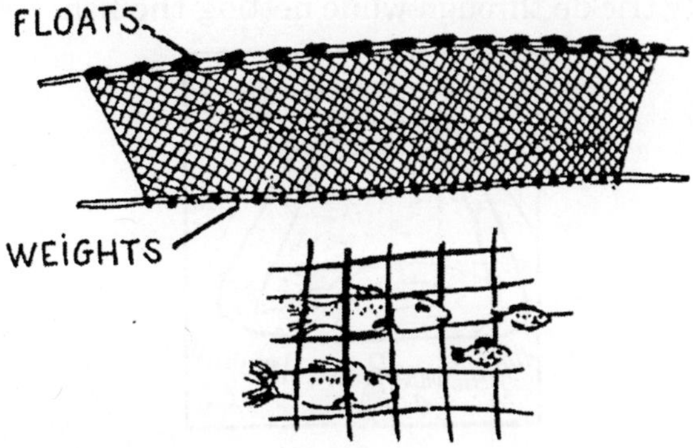

They are called gill nets because the fish pokes his head through the net mesh, and is caught around the gills as he tries to wiggle through the net. Another net used to harvest fish is the seine. **A** seine can collect all the fish in the pond at one time because it has smaller openings (mesh size) than the gill nets, and it is usually made of heavier fibers to hold the fish. (See the end of this section for instructions on making a seine).

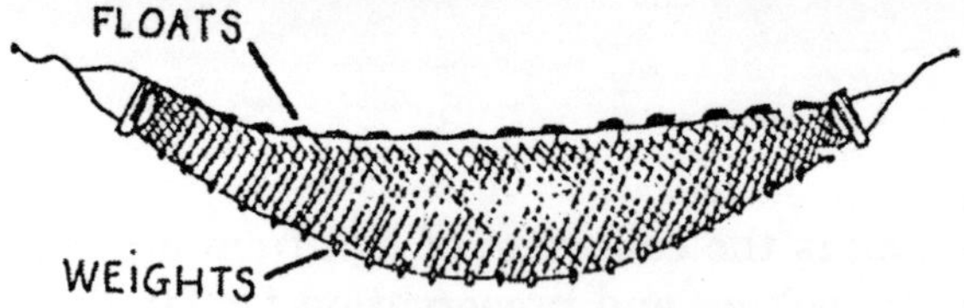

Both seines and gill nets have lead sinkers (weights) attached to the bottom ropes. These weights hold the nets at the bottom of the pond (so the fish cannot escape underneath the nets as they are pulled). Seines and gill nets also have floats attached to the top ropes to help the net form an enclosure: the entire pond is netted with one sweep of the net.

Netting a Pond

Let out as much water as possible. NEVER LET THE WATER OUT COMPLETELY. As the fish have less and less water in which to live, they become excited and use up more oxygen when there is less available. Plan on harvesting while the water is draining out so the fish are caught before they are stressed. Or, drain the pond almost completely, and then let water slowly trickle through while netting the fish.

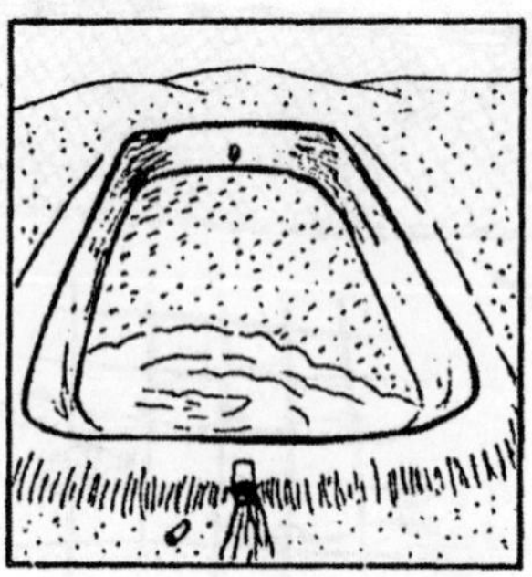

USING A SEINE Place the net at one end of the pond and slowly draw the edges down the sides of the pond. Bring the middle of the net across the pond. When near the other side, begin pulling the edges up onto the bank so that the net forms a u-shape in the pond. Pull up the bottom rope of the net along the pond bottom until it breaks the water surface. At this point the net is a bag shape and will hold the fish in (some seines already have a bag woven into them).

Pick the fish up one by one and transfer them to buckets or tubs of clear water for later weighing and transport.

OTHER HARVESTING METHODS Fish also can be harvested by other methods. One method is to catch them with a hook and a line, but this method is time-consuming. In some parts of the world fish are harvested by dynamiting or poisoning the water. But these methods are dangerous and should never be done in a pond or any other waterway: dynamite and poisons can kill people and other animals, in addition to fish. NEVER HARVEST FISH BY DYNAMITING OR POISONING THE POND. There are easier and cheaper methods than these.

Marketing Harvested Fish

Once fish are harvested, they must be marketed. Marketing includes the transportation and sale of fish. As the introduction to the manual pointed out, one very important thing to consider before building a pond is the availability of a market. If a market is further away, the farmer must have transportation to it over passable roads. If the market is very near, he may want to advertise the date of his harvest by word-of-mouth so that the people will come directly to the pond to buy the fish. Also, he may want to make an agreement with a tradesman at the market so that he is sure he has a buyer for his fish when they are harvested. If there is no market, or if the farmer is going to use all the fish himself, then he probably will want to preserve some of the fish (see fish preservation).

Transporting fresh fish to market must be carefully done, so that the fish are not damaged. Usually, fish are handled in the same way they were handled when put into the pond. If it is not possible to get the fish to market right away, they must be preserved -- either on ice for quick sale in a nearby market; or salted, dried, smoked, or canned if going to a distant market. These methods are discussed in the next section.

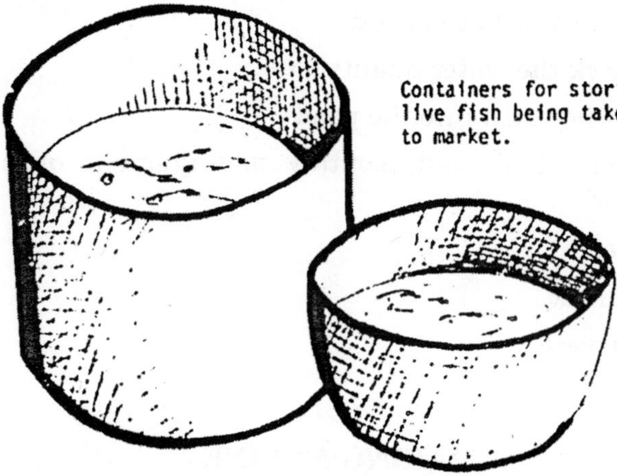

Containers for storing live fish being taken to market.

Remember: Fish spoil very quickly in warm temperatures. Sell or preserve the fish right after harvesting.

After Harvest

After the pond is harvested, it should be prepared for the next stocking of fish:

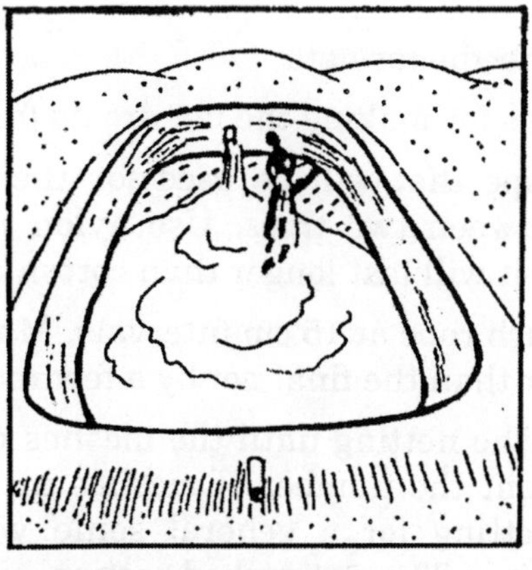

- Plow the bottom of the pond
- Clear out predators, sticks, rocks, etc.

- Dry the pond bottom until the soil cracks
- Put lime on the pond bottom
- Wait two weeks
- Add water to the pond
- Check the water quality
- Put new fish into the pond
- Begin daily and monthly management of fish and ponds
- Breed
- Market
- Harvest
- Begin again

MAKING A SEINE

A seine can be made using materials found in the market. The materials needed are:

- rope
- cork floats
- lead sinkers (or something heavy to help the net sink)
- netting
- sewing needle for nets

The directions for making the net are as follows:

- Tie a rope that will be used for the top and bottom lines between two trees. Use nylon rope, if possible, because it will last longer than cotton or hemp.
- Mark each rope at 15 cm intervals. Make sure the rope is longer than the final net by a few meters.
- Stretch the netting until the meshes close completely; then count the number of meshes in a 23 cm section. Good netting for a general seine will have 6 to 9 meshes in a 23 cm stretched section.

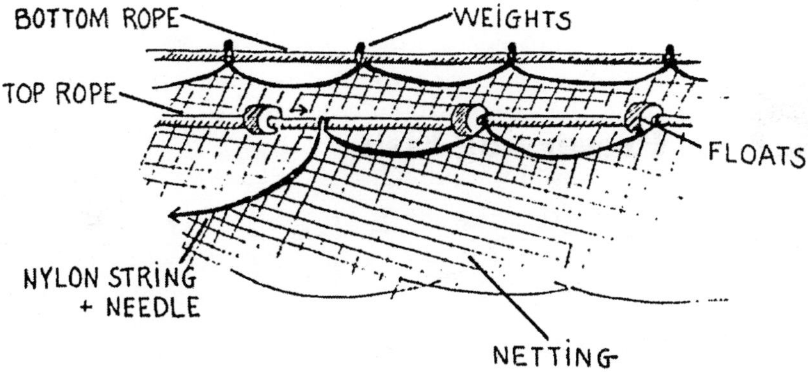

- Use nylon string that is very strong. Wind a long section on a net needle. Then tie the end onto the lead line rope (top rope) at the first marking. Pass the needle through the number of meshes counted in the 23 cm section of netting. Tie the string on the rope at the second marking.
- Repeat the process until the last marking on the top rope is reached.
- Pound the sinkers, or string them, onto the bottom rope at the 15 cm intervals. Tie the cork floats onto the top rope at the same intervals.
- String the bottom line onto the netting in the same way as the top line.

REMEMBER: The net must be washed, repaired, dried in the shade, folded, and put away in a cool, dry place after each use. A net which is taken care of in this way will last much longer.

8 Preserving Fish

Fish that are not taken to the market fresh must be preserved in some way after harvesting. All fish have bacteria in their intestines; as soon as they die, these bacteria begin to multiply and the process of decay begins. So the first thing which must be done -- as soon as possible -- is to remove the intestines. After this is done, go on to preserve the fish in the way chosen.

There are a number of ways to preserve fish: salting and smoking are discussed here in some detail.

Salting Fish

Salting is a very old method of preserving fish. Salting depends on the size of the fish, the species, and on the amount and quality if the salt used. Fish which have been salted well last a long time without spoiling.

The most important factor in salting fish is the quality of the fish being salted. Use only fresh fish: fish which have been lying around for hours are not good for salting. Also, use only clean equipment end clean fish.

PLEASE READ THE DIRECTIONS THROUGH CAREFULLY BEFORE BEGINNING.

TOOLS AND MATERIALS

- Clean sharp knife
- Salt -- about 20 kg for each 100 kg of fish

- Containers for washing fish (buckets, tubs, drums)
- Flat working surface (table, flat stones)
- Containers for holding waste (parts of the fish not used)

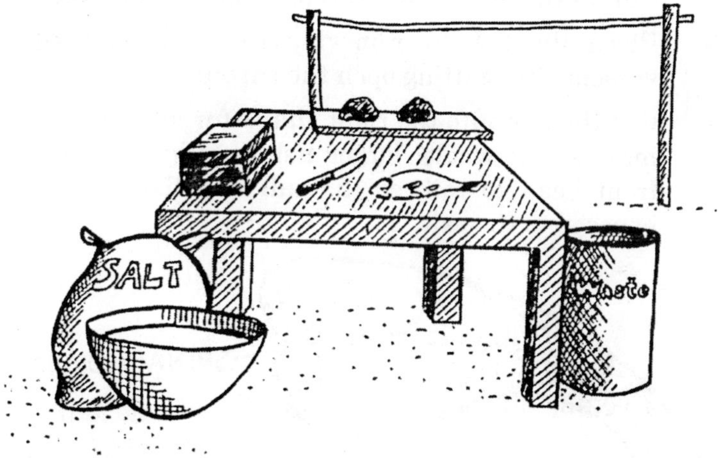

- Waterproof boxes or jars to hold salted fish (glass or wood; not metal unless the metal is stainless steel) -
- Boards and weights (to press down the fish)
- Slats or lines for drying the fish
- Small shelter to cover fish while drying

STEPS IN SALTING FISH

There are four major steps: gutting and cleaning; salting; washing and drying to remove excess salt; and, finally, air drying.

Gut and Clean the Fish.

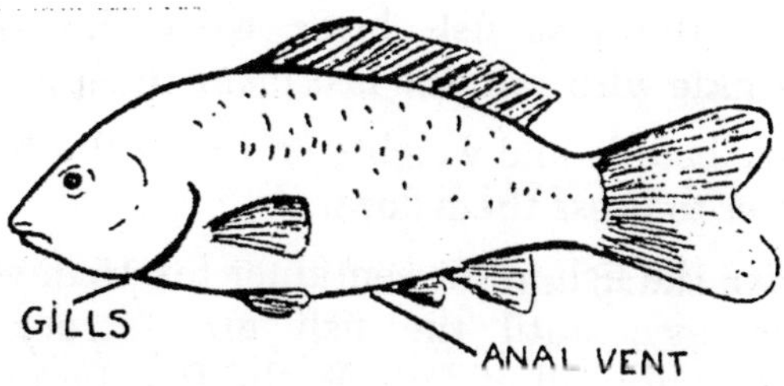

- Gut the fish by cutting along the belly from the gills to the anal vent.
- Remove the guts and the black membrane in the gut cavity.
- Cut off the head now, if preferred; it is not necessary.
- Bleed the fish by removing the gills and all blood vessels after cutting open the throat.
- Cut the fish into the right shape for salting: small fish may be left whole; larger fish should be split in half from head to tail, so that all the fish flesh will be exposed to the salt.

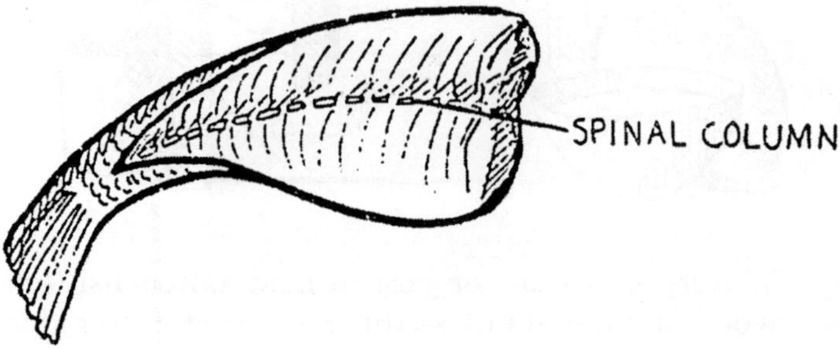

Salt the Fish.

- Sprinkle a layer of salt on the bottom of the container which will hold the fish.
- Place a layer of fish, flesh side up, on the salt. Do not let the fish lay on top of each other.
- Cover the fish with a thin layer of salt.
- Continue to place fish, then salt, almost to the top of the container.
- Place the last fish layer with the skin side up. Sprinkle with salt; the last layer must be salt.
- Place boards and weights on top of the fish in the container to press them down.
- Leave the fish in the container for 15 days. Add salt as necessary, until the fish are "struck through" -- thoroughly full of salt. As the fish lie in the salt, the

salt draws out all the moisture in their flesh. This moisture forms a solution (brine) with the salt as the salt dissolves. It is necessary to add more salt as the salt is diluted in the solution. As the moisture is removed from the fish by the salt, the level of fish in the container falls.

- Add more fish, skin side up, and also more layers of salt as the level of fish falls.

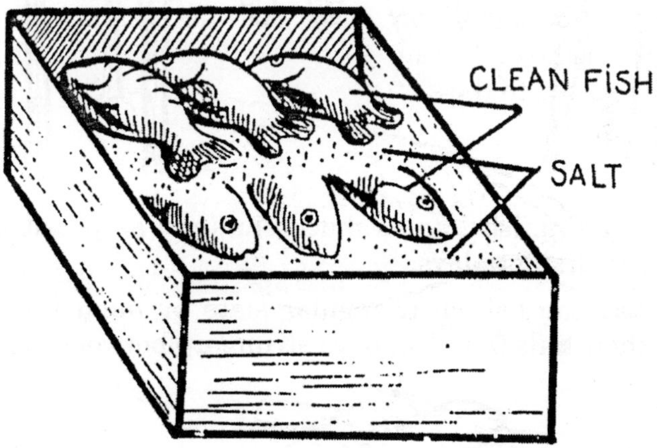

Wash and Dry the Fish.

- Remove the fish from the container when they are fully salted. The fish are properly salted when they are firm and have a whitish salt layer on their flesh.
- Wash the fish in clear, clean, sea water or brine.
- Place the fish on a flat surface and press them down with boards and weights to make them as flat as possible before drying.

Air Dry the Fish.

- Dry the fish in the sun and in the air, or use heating and fans. Usually fish are dried outside in an area that is exposed to sun and wind and is very clean.

- Dry the fish under a shelter of leaves or branches for the first few days, so that they do not dry too quickly.

- Put the fish into as much sunlight as possible, after the first few days.
- Lay the fish on triangular slats or hang the fish by their tails from fish lines strung up between trees.

- Cover the fish if it rains. Any moisture at all, at this stage in the salting process, will cause the fish to spoil.
- Dry the fish for about six days.
- Pack and store the fish in water proof containers.

HOW TO USE SALTED FISH

Soak salted fish in fresh water overnight. Change the water at least or during this time. The soaking removes the salt; the longer the fish is soaked, the more salt is removed. After the fish has been soaked, it can be used in any way that fresh fish is used.

Smoking Fish

Smoked fish does not last as long as salted fish, because it must be refrigerated, frozen, or canned if it is to be stored. Smoked fish are prepared in a smoke house which is merely a shed or a box over a fire which is controlled so that it produces smoke instead of flames. The fish are merely hung inside the smokehouse so that they are surrounded by smoke. It takes about six hours to smoke fish so that they can be eaten or stored.

Smoked fish are prepared like fish for salting. After they are bled, and gutted, they are split from head to tail. They are then washed in freshwater and placed in a saltwater brine made by dissolving 1 kg of salt in one liter of water for one hour. Then the fish are removed from the brine and washed in clean, freshwater again. The fish are then drained and hung in a cool breezy place for about an hour.

At this point, the fire can be built in the smokehouse. When it is smoking properly, place the fish on hooks and hang (or tie) the fish in the top of the smokehouse. Make sure the fish are placed securely **so** they will not fall. Watch the fire carefully to make sure it is smoking, and not burning, the fish.

After the fish are smoked for six hours, they can be eaten immediately, or stored in jars (to be canned), or stored from zen or refrigerated until they are eaten.

Smoked fish do not last as long as salted fish, so do not smoke all of the fish, unless it will be used soon after harvesting.

Other Preserving Methods

Fish can also be preserved by simple air drying, or by canning. Air drying involves only cleaning and washing the fish and drying them in the sun and wind until they are a clear white color. Canning is a much more complicated process. Canning must be done very carefully: fish can contain many bacteria which must be killed before canning. If fish are canned with this bacteria still in them, the fish will spoil. People who eat canned fish which is spoiled can become very sick. A farmer who wishes to can his fish should arrange with a canning factory to take part of his harvest and can it for him.' A farmer should not try to can fish at home unless he has expert help.

Often fish are preserved by freezing. Freezing requires a constant supply of electricity -- which most farmers do not have. If electricity is available, however, freezing is one of the easiest and safest ways to preserve fish. In this method, the fish are gutted, cleaned, cut up (if desired), placed into containers, and put into freezers. Frozen fish can last for a very long time, if they are not thawed (unfrozen). Once frozen fish are thawed, they must be used immediately, or they will spoil.

Spoiled Fish

Even spoiled fish can be used --although it cannot be eaten by human beings. Spoiled fish can be cut up and boiled, then dried in the sun or cooked in an oven until it is very flaky. Once this is done, grind the fish into a powder and mix it with powders of plants: this makes a very nutritious food for fish in ponds. The powder can be used as a powder, or it can be mixed with something to make it stick together so that the powder can be pressed into pellets for fish.

Spoiled fish, and even the guts of fish that have been used in some other way, are called "trash" fish. The powder is called "fish meal." Fish meal is used to feed fingerlings or even brood stock. Fish meal is one of the best fish foods for pond fish.

9 Problems of Fish in Ponds

Fish cultured in ponds can have problems: they can be stressed by a lack of oxygen; they can be eaten by predators; they can be infested by parasites. These problems and some solutions to these problems are discussed in this section.

Diseases

Diseases of pond fish are caused by fungi, bacteria, protozoans, worms, and crustaceans. Usually diseases can be controlled by proper pond management, which includes draining the pond, drying it, and liming it periodically, and also by preventing wild fish or unfiltered water from entering the pond. Some diseases are fatal, but many can be controlled by treating the pond or the fish with chemicals.

Some diseases attack fish in ponds because some other factor is causing stress: overcrowding, low oxygen levels, or not enough food. All of these conditions weaken the fish so they can get diseases more easily. The farmer must watch his fish for signs of stress and disease. Any change in normal behavior may be a sign of disease; for example, gasping at the surface for air, rubbing the body or head against the sides of the pond, or ragged fins and sores on the body. Something is wrong when a fish population stops eating suddenly. So the farmer must check the fish often (see "Management"), especially in very hot weather.

FUNGAI DISEASES These diseases are caused by fungi.

Gill Rot. This is a disease caused by the filamentous fungus, *Branchiomyces sanguinis*, This disease is first noticed by a red spotting of the gills. Later, the gills become greyish-white and stop working. When the gills stop working, the fish suffocate and die. Gill rot is most common during the hot part of the year and is sometimes associated with large amounts of dung and a "bloom" of plankton.

Treatment: Remove dead fish from the pond; the remaining fish will probably recover. Drain the pond and dry the bottom. Treat the pond with quicklime or copper sulphate to kill the fungus spores. Fill the pond again. Add quicklime every few weeks until there is no more sign of the disease.

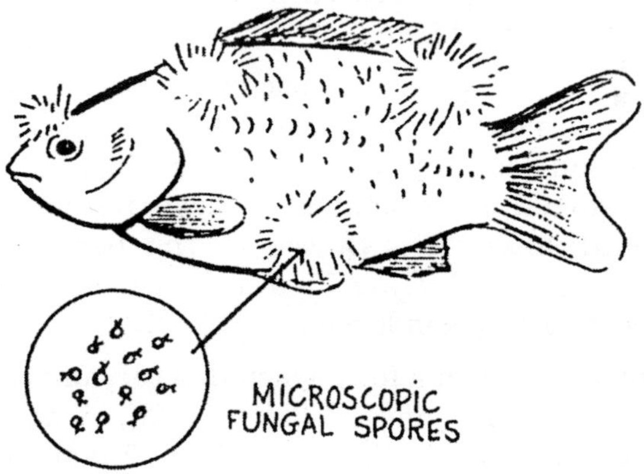

Saprolegnia. This fungus is often associated with Gill Rot. It attacks weakened places (e.g., bruises from handling) on fish. Since it hits already weakened fish, *Saprolegina* attacks fish already trying to fight other diseases. *Saprolegina* looks like fuzzy, white cotton wool and is often in tufts on the body of the fish. *Saprolegina* by itself can kill eggs and fry, but does not kill adult fish. Indian carp are very susceptible to this disease, and common carp eggs are attacked frequently.

Treatment: Use the same treatment as outlined for Gill Rot.

BACTERIAL DISEASES These are caused by parasites which are actually bacteria.

Furunculosis. This is the most important bacterial disease. This disease causes ulcers or abcesses in muscle tissue. It then breaks through the skin, and, eventually, becomes a site for fungus infections, like *Saprolegnia*. This disease attacks in the spring, and is most often found in more temperate species, like trout.

Treatment: Drain the pond and treat it with slaked lime. Disinfect every tool used in the pond (nets, feeding rings, etc.).

Infectious Dropsy. This is caused by the bacterium, *Pseudomonas punctata*. The symptoms are a swelling of the fishes' belly with water, ulcers on the skin, lengthening of the fins, and deformation of the backbone.

Treatment: Prevent diseased fish from entering the pond. Bury and burn the dead fish.

Columnaris. This is another bacterial disease which causes discolored patches on the body, loss of scales, and, often, death. This disease can look like a fungal disease, but it is not. If possible, it should be examined under the microscope for positive identification. It is caused by the bacteria *Chondrococcus columnaris* and *Cytophaga columnaris* and is often associated with low oxygen leveis.

Treatment: Give fish a feed which has terramycin in it. If it is very bad, place each infected fish in a dip (bath) of copper sulfate (2 minutes in a solution of 1 to 2,000) or a dip of malachite green (10 to 30 seconds in a solution of 1 to 15,000). Treat the pond with 1 ppm of copper sulfate.

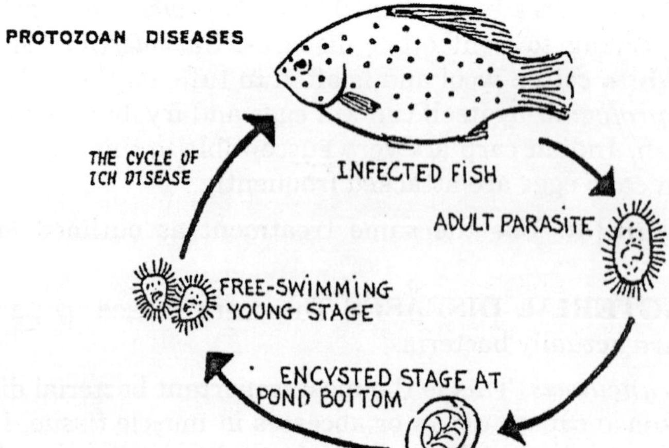

Ichthyophthirius multifilis. This is the worst protozoan disease. The "ich" disease is caused by a ciliate which forms white spots or pimples on the skin and fins of the fish. Each parasite produces thousands of spores, which can then infect other fish in the pond.

Treatment: Drain the pond, and lime it. Or treat the fish with chemicals as follows:

Formalin	200-250 ppm	daily bath
	15 ppm	in pond
Malachite green	1.25 ppm	daily bath/30 minutes
	0.5 ppm	in pond
Methylene blue	2 ppm	daily bath
Acriflavin	10 ppm	3-20 daily baths
Salt	7,000 ppm	several daily baths

Costia and *Trichodina*. These are two other ciliate diseases. They are caused by microscopic organisms which attack the skin of fish and cause lesions. Tilapia, the very resistant fish, are attacked by the *Trichodina* protozoan.

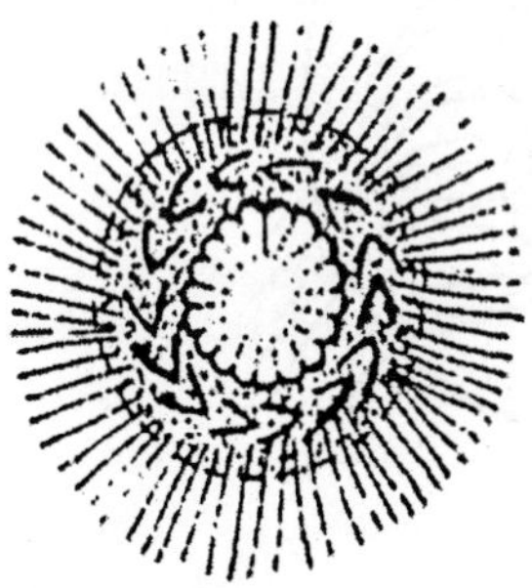

TRICHODINA

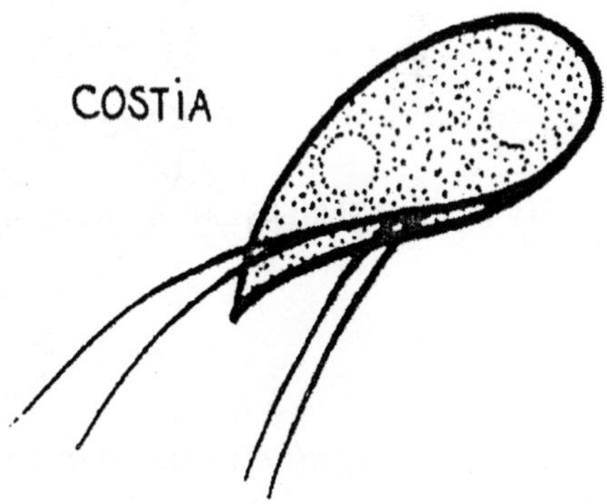

These ciliates cannot be seen by the naked eye, but the lesions and sores that they cause can be seen by looking closely at the fish.

Treatment: Add 3 ppm of potassium permanganate to pond. Or dip the fish in baths of 5 to 10% sodium chloride (salt) for 5 to 20 minutes daily for up to one week.

CRUSTACEAN PARASITES

Lernea. The anchor worm is the most common disease of this type (accepted). This worm attacks the gills or any other part of the body. It burrows into the fish, leaving its two eggs cases protruding on the outside of the fish. *Lernea* causes red sores, and makes the fish thin so that their market value is much lower.

LERNEA

Treatment: Add castor oil in a thin film over the surface of the pond. Treat fish infected with young *Lernea* in a formalin bath, or remove each parasite by hand.

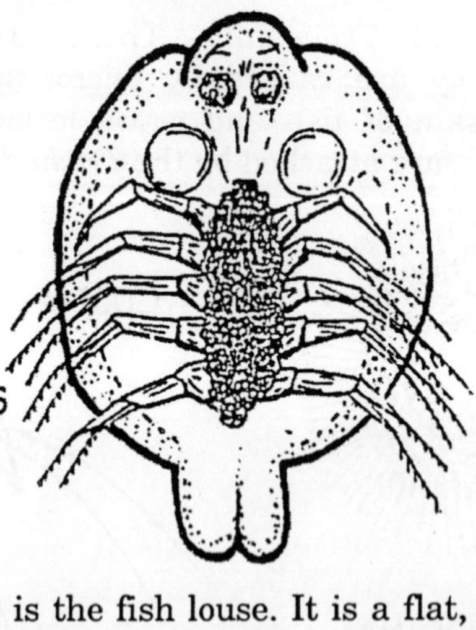

ARGULUS

Argulus. Argulus is the fish louse. It is a flat, pinkish-red disc that clings to the skin, fins, mouth, or gills. It sucks blood with a piercing organ, which also injects poisons. Young fish may die.

Treatment: Drain and lime the pond. Or place the fish in a bath of 3 to 5% salt, or 250 ppm of formalin for 1 hour.

WORM PARASITES - Most of these are external parasites.

Dactylogyrus. This parasite attacks the gills of young fish. The fish are exposed to this worm when they are between 2 and 5 cm long.

Treatment: Manage the pond well so that fingerlings grow rapidly past the stage when they are susceptible to *Dactylogyrus.*

Gyrodactylus. This parasite burrows into the blood vessels of fish through the skin, causing the fish to appear reddish with sores. This worm can cause fish to die from emaciation.

Treatment: Treat ponds with 5 ppm formalin. Treat fish individually in a bath of 25 ppm formalin.

Bothriocephalus gowkongensis. This is the tapeworm which often attacks the Chinese carp, especially grass carp. It is difficult to treat this worm; it is found in the fishes intestines.

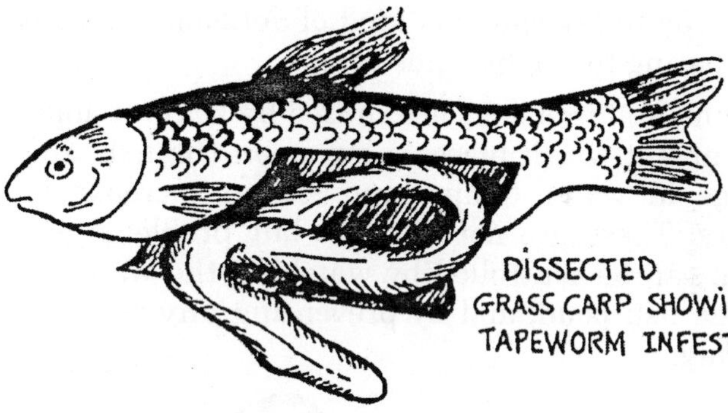

DISSECTED GRASS CARP SHOWING TAPEWORM INFESTATION

General Treatments

Farmers often will have trouble finding the proper chemicals for treating their ponds or deciding which disease the fish have and which treatment to give. Here are some general treatments: any of these treatments will help an infected pond.

Baths:

Potassium permanganate	4ppm
Salt	3-5%
Copper Sulfate	500 ppm for 1-2 minutes
Formalin	250ppm for 1 hour

Malachite green 67ppm for 10-30 seconds

Or the farmer can use unslaked lime directly in the pond.

Some pond owners always treat new brood stock with a one-hour bath in 10 ppm of potassium permanganate, and then transfer the fish to a bath of 15 ppm of formalin for 4 to 12 hours. This ensures that no parasites will be introduced into the pond with the brood stock.

Other Problems

Other problems are caused by deficiency or environmental factors.

Deficiency problems appear because the fish are missing some factor they need to grow and be healthy. The missing factor can be a lack of essential elements like vitamins or minerals. This lack is difficult to detect until a problem exists. So the only way to prevent this kind of deficiency is to be sure the fish are eating the right kinds of food.

Environmental problems are caused by some change in the pond environment which places a stress on the fish, such as a rapid change in water temperature or an increase in pond water acidity. These are not diseases but problems of fish in ponds which can be controlled by watching the water and soil quality of the fish pond, and by preventing any rapid changes from occurring.

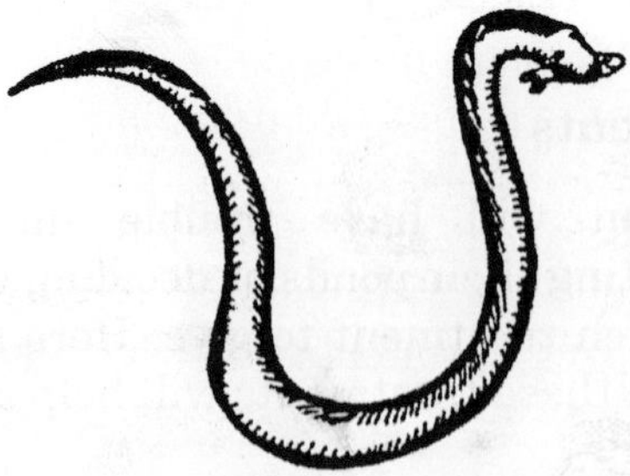

Predators

Other problems occur in fish ponds when other animals eat the fish. Frogs, snakes, and birds eat young fish and must be kept out of ponds. The worst predators, of course, are carnivorous fishes, like the *Clerias* catfishes. Prevent these fish from entering the ponds by screening the water inlet.

In any pond, all unwanted (trash) fish and predators must be removed before stocking the pond. If the pond can be emptied, simply drain the pond, plow and dry the bottom, etc. If the pond cannot be drained, seine the pond as completely as possible. However, many fish escape the net by staying at the edges of the pond. The best way to get rid of the predators is to poison the pond water in a pond which cannot be drained.

USING POISON The most common poison for use in fish ponds is rotenone. Rotenone can be purchased --as a liquid or powder --or it can be gotten from the roots of the derris plant. To make rotenone, collect derris roots and pound them until a milky-white fluid can be squeezed out. This fluid contains rotenone. Apply one kilogram of derris root for every hectare of pond surface area. If using powdered rotenone, use only 0.05 kg/ha. The powder should be dissolved in water and dipped into the pond from buckets.

Other poisons used in fish ponds are quicklime, tea seed cake, camelia seed cake, tobacco waste, and powdered croton seed. These are some application rates:

Quicklime:	160kg/ha
Tea seed Cake:	150kg/ha
Camelia Seed Cake:	50 to 200kg/ha, depending on depth
Powdered Croton Seed	
Seed:	50 to 200kg/ha, depending on depth
Tobacco Waste:	150 to 200kg/ha

Most of these natural poisons will degrade (break down) and disappear from the water in 7 to 12 days. After this period, seine the pond again. If no live fish are caught, stock the pond.

There are many chemicals which can be used to poison predators in fish ponds. However, many of them stay in the ground too long. Others are dangerous. One of these chemicals which can be used safely is saponin, which is a component of Tea seed Cake. Apply a dose of 0.5 ppm in the pond.

In most places, there are fishermen and farmers who know of some local plant which causes fish to die. For example, in India large ponds that cannot be drained are poisoned with Mahuca oil cake (*Mahuca latifolia,* syn. *Bassia latifolia*), applied at a rate of 150 to 250 ppm (1500 to 2500 kg/ha per meter of water depth). This plant poison breaks down in 10 to 20 days. These types of poison are all better sources of poison than chemicals. Many times, when there is a tree that overhangs a pond, fish will be killed when the tree leaves drop into the pond. Watch for plants which do this, and use them in ponds instead of poisons in a chemical form.

DO NOT USE CHEMICALS LIKE ENDRIN, DIELDRIN, AND DOT IN PONDS: THEY CAN LAST IN THE GROUND FOR YEARS, AND LATER, KILL THE POND FISH. NEVER USE POISONS WITHOUT FIRST CHECKING WHETHER THEY CAN BE USED IN PONDS. POISONS KILL OTHER ANIMALS AND HUMAN BEINGS, AS WELL AS FISH.

SUMMARY : FISH DISEASES & TREATMENTS

DISEASE	DISEASE ORGANISM	TREATMENT IN PONDS	TREATMENT IN BATHS
Bill Rot	*Branchiomyces sanguinis*	Quicklime Copper Sulfate	
Saprolegnia	*Saprolegnia*	Quicklime Copper Sulfate	
Furunculosis		Slaked lime in Drained Pond	
Infectious Dropsy	*Pseudomonas punctata*	Burn or Bury Dead Fish	
Columnaris	*Chondrococcus columnaris* *Cytophaga columnaris*	Copper Sulfate 1ppm	Copper Sulfate 500ppm for 2 minutes Malachite Green 67 ppm for 10-30 seconds

Ich	*Ichthyophthirius multifilis*	Formalin 15 ppm Malachite Green 0.5 ppm	Formalin 200-250 ppm Malachite Green 1.25 ppm Methylene Blue 2 ppm Salt 7000ppm Acriflavin 10ppm
Costiasis and Trichodiniasis	*Costia* and *Trichodina*	Potassium Permanganate 3 ppm	Salt 5-105 for 5-10 minutes daily
Anchor Worm	*Lernea*	Castor Oil	Formalin
Fish Louse	*Argulus*		Salt 3-5% Formalin 250 ppm for 1 hour
Nematodes	*Dactylogyrus* and *Gyrodactylus*	Formalin 5 ppm	Formalin 25ppm

10 Other Methods of Fish Culture

Fish culture in ponds is the primary method of freshwater fish culture. However, there are other methods of fish culture used in places where ponds are not possible.

Fish Culture in Dams and Reservoirs

Water contained by dams and reservoirs is sometimes used for fish culture. These waters can be stocked with fry or fingerlings; the adults are later harvested with nets. Raising fish in these waters is more difficult than in ponds because these waters cannot be drained, and the predators cannot be removed. Also, it is not possible to feed, fertilize, or poison the water, so natural nutrients must provide enough fish food. But if there is no other water source available, culture in dams and reservoirs can work.

Culturing fish in waters held by dams and reservoirs can be done more easily if the fish are placed in fish cages and pens. These structures confine the fish to a certain place and give more control over the fish.

In Cages

In many parts of the world, the only water available is flowing water or large bodies of water where it is not possible to divert the water into a pond. In these waters, it is possible to grow fish in small cages. Cage culture can also be practiced in areas like swamps where there is water not being used for any other purpose.

Cages can be rectangular boxes, bamboo cylinders, or anything that can be floated in a water current so that the water passes through.

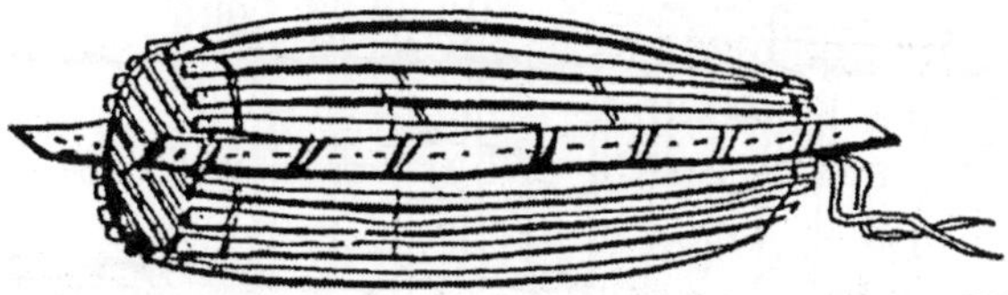

In addition to bamboo, cages can be made out of such materials as wire screen, nylon mesh, and wood. All cages must be anchored so that they do not float away.

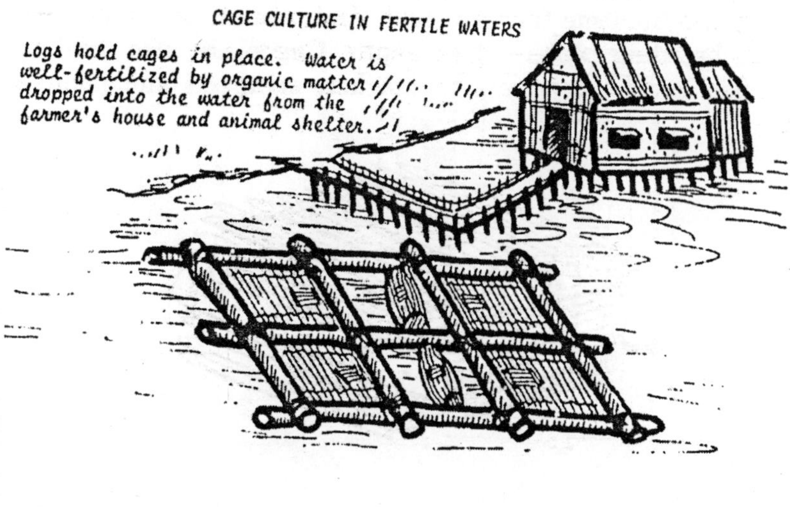

Cage culture is used in some countries in very fertile waters (polluted from sewage) with very good results. Fish in cages usually get their food from the water as it floats past the stationary cage, but in some cases, the caged fish are fed pellets of food daily.

Fast flowing water is best for cage culture. If the water is not flowing very fast, problems such as oxygen lack and competition for food can occur. These can be big problems in cages because there are usually more fish placed in the small area of the cage than would normally be in the same area in the pond.

Cage culture is still experimental, but in ideal conditions, good growth rates have been shown by fish that were grown in cages and given extra food.

Cages also are used inside ponds for holding fish between harvest and the time they are sold. And, sometimes, cages are used as breeding tanks -- like hapas. Cages are also used to carry fish caught in rivers to market, strapped alongside a boat.

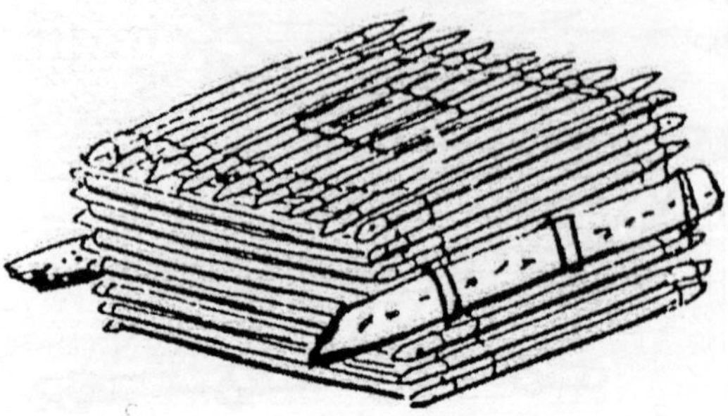

In Pens

Fish can also be cultured in pens inside lakes or offshore areas. Fish culture in pens has been done in Israel and Scotland for years, and is now being done in some Asian countries. Pens are constructed of bamboo or wooden poles that are forced down into the lake or shore bottom. Then nets are strung from pole to pole to form an enclosure. The nets are anchored into the lake bottom with weights or sinkers, and the fish are placed inside the pen for culture. Fish grown in pens can be larger controlled a little better than fish in cages because pens are (fish pens can be comparable in size to regular fish ponds) and provide more area and more food.

Fish pens placed in fertile (productive) lakes have very good growth rates. In a fish pen placed in a major lake in the Philippines, silver carp stocked at 7 grams gained an average of 4 grams a day in a 52-day growing season.

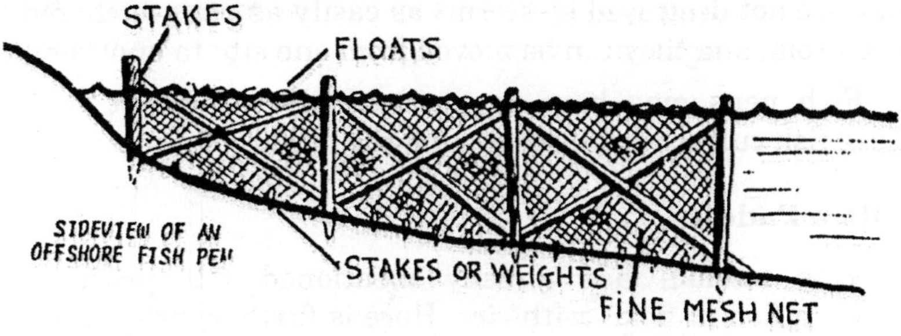

Fish pens have many good points: they require no extra feeding of fish, no fertilization, and very little maintenance (although a lot of care is given to the nets). The fish are stocked and harvested later at the end of their growing season. Fish pens can work in water is areas where the not very productive, but in these areas, the fish must be fed supplementary foods. Feeding rings are used so the food will stay in the pen and not float out into the water. harvested by gill nets; Fish in pens are usually seines also may be used.

There are some disadvantages to pens:

- Pens are expensive to build. The netting used must be nylon or plastic so it does not rot, and poles must be treated so they do not become waterlogged and rot. In the Philippines, it costs about $1,428 (U.S.) to build a one hectare pen, using nylon netting and bamboo poles. This is comparable to the cost of a one-hectare fish pond, but a pen can be destroyed by a big storm and a pond will not be destroyed.
- A fish pen only lasts three to five years in the water.
- Fish pens are usually built in the shallow areas of a lake, where they use space many fish need to feed and spawn. The pens, therefore, reduce the natural production in some lakes.
- Fishermen must go further out into the water to fish when pens are in the shallow areas.

Fish pens can also be built like fish cages so that they float. Floating fish pens are used most for marine fish research studies; they also can be used in lakes. Floating fish pens can be as small as one hectare in size, or as large as 10 hectares. They are not destroyed by storms as easily as pens anchored to the bottom, and they can be moved from one site to another.

Fish pens may have an increasingly important role in future fish culture activities around the world.

In Rice Paddies

This manual has already mentioned the practice of culturing fish in fields with rice. Here is further, brief mention of that subject. The farmer digs deep trenches all along the dikes of the paddy. He then floods the field and plants the rice. After the rice has grown to a height of 5cm or so, fish can be placed into the paddy field.

This culture method can be used only with fish that are resistant to low oxygen levels and are not herbivores -herbivores might eat the young rice plants. *Clarias* catfishes are good fish to culture in rice paddies because they have accessory breathing organs which help them to breathe even when the paddy gets dry and the water in the trenches gets very low.

After the rice is harvested, the fish are caught in hand nets and sold. This is not really a culture of fish, but a culture of rice with some fish added. It can be an easy way for a farmer who has no extra land on which to build fish ponds to increase the total production of his land.

Glossary

acclimate - to become adjusted to a change from the normal environment (also *acclimatrize*).

acid - a substance that can dissolve in water and is sour or bitter in taste, and turns litmus paper from blue to red.

adhesive - a sticky substance; sticking or sticky to something else.

aerotion - adding oxygen to water by spraying or bubbling air through the water.

algae -small or large water plants from five classes of plants.

alkalinity - the ability to combine with an acid to form a salt.

aquaculture - the cultivation of animal and vegetable life in water.

area - the length times the width of a piece of land or other surface.

back washing - forcing water in the opposite direction from its normal flow.

barbels - sensitive organs that hang down on the sides of the mouth of certain fishes.

basic - having base forming elements (alkaline on reaction).

bloom - a very good growth of algae in a pond that has a strong green color.

bottom feeders - fish that feed on bottom organisms (organisms that live in mud on the pond bottom).

breeding - the cycle of reproduction in animals.

brine - water that is saturated with common salt, or the water from a salt water body (the ocean).

brood ponds - ponds where the fish used for breeding are kept.

brood stock - the fish used for breeding in fish ponds.

cage - an enclosure to hold fish in the water.

captivity - the state of being held in a confined place (fish in ponds are captive).

carnivore - an organism that eats animal products.

centrifuge - the machine that uses centrifugal force to separate materials of different densities.

compete - to fight for something against someone or thing.

contaminant - something that makes something else impure; a pollutant.

cooperative - an organization of people that are working together for a common purpose.

dam - the wall of a fish pond.

debris - rubbish, garbage, anything that is not supposed to be in a certain area (pond).

density - the number of fish in a pond.

dike - the wall of a fish pond.

diversion channel - a ditch that takes water from a stream or river to a fish pond.

elevation - the height of land.

exotic species - fish cultured in ponds that are not native to the area.

fertility - being very productive.

fertilizer - anything added to water or soil to make it more productive.

fingerling - a fish that is about as long as a man's finger (6-10cm).

fishculture - the breeding and cultivation of fish in ponds.

fry - fish that have just hatched until they reach fingerling size.

genitals - reproductive organs.

genital opening - -the opening on the fishes' body where the eggs or sperm are released.

gills - the part of a fish that allows it to breathe in the water.

gravity - the tendency of things to fall downwards towards the center of the earth.

hapa - the mesh enclosure in ponds where fish can be spawned.

herbivore - an organism that eats only plants and plant products.

hypophysation - hormone injection to induce breeding of fish.

hypophysis - the pituitary gland.

hormones - components that are secreted by glands of the body to cause certain changes in the body's functions.

impermeable - a substance that nothing can leak thru.

induced spawning - causing a fish to spawn by injecting it with hormones.

introduced species - fish not native to an area that are used in fish ponds of the area.

kakaban -an egg collector.

mortality rate - the rate of death.

natural food - food that a fish eats in nature.

niche - what an organism does; its job in the community.

nutrient - an ingredient of food-that is healthful.

omnivore - an organism (like man) that can eat both plants and animals.

operculum - the gill covering.

oxygen - a gas that is necessary for all life.

pens - enclosures for fish culture on large bodies of water.

phytoplankton - tiny green or brown plants that are microscopic, free-floating in water, that are used as food by fish.

photosynthesis - the process on which green plants produce food for themselves and release oxygen into the water.

pituitary gland - the gland that releases hormones controlling the reproductive cycle in animals (like fish).

plankton - the tiny plants and animals that grow in ponds that are eaten by fish.

ponds - any enclosure that holds water so that fish can be grown inside it.

predators - animals that prey on other animals.

productivity - ability to grow food in a pond, whether it is plankton or fish.

reproduction - producing offspring.

respiration - breathing.

serrations - rough edges, like on a fishes' fin.

slope - the slant of land.

spawning - the release and fertilization of eggs and sperm.

stress - any change that is not normal in the environment that creates problems.

trash fish - fish not wanted in the pond, *or* fish that are too small to eat *or* spoiled fish.

watertight - impermeable.

zooplankton - small animals in ponds that can be seen with the naked eye.

Resources

American Public Health Association. 1971. Standard methods for examination of water and wastewater. 13th ed. Am. Pub. Health Assoc., Washington, D.C. 874 p.

Anderson, Steven E. 1973. A manual of fish farming for tropical Africa. University of Minnesota, St. Paul, Minn. 46 p. (xeroxed copy)

Avault, James W., Jr., 1965. Preliminary studies with grass carp for aquatic weed control. The Progressive Fish Culturist. 27 (4): 207-209.

Avault, James W., Jr., and E.W. Shell. 1966. Preliminary studies with the hybrid tilapia *Tilapia nilotica* X *Tilapia mossambica*. FAO World Symposium on Warm Water Pond Fish Culture. Rome, Italy.

Avault, James W., Jr., R.O. Smitherman, and E.W. Shell. 1966. Evaluation of eight species of fish for aquatic weed control. FAO World Symposium on Warm Water Pond Fish Culture. Rome, Italy.

Aylward, Francis and Mogens Jul. 1975. Protein and nutrition policy in low-income countries. Charles Knight and Company, Ltd., London. 150 p.

Bardach, John E., John H. Ryther, and William O. McLarney. 1972. Aquaculture. John Wiley & Sons, Inc., New York. 868 p.

Beckert, Heino. 1967. Culture of some common fish parasites for experimental studies. Zoology-Entomology Dept. Series, Fisheries 5. Agricultural Experiment Station, Auburn University, Auburn Alabama. 28 p.

Best, Cody D. 1975. personal communication

Bharadwaj, R.S., Stephen Crawford, and Lauren C. Watson. 1973. Manual for fish culture in Rajasthan and Madhya Pradesh. American Peace Corps. New Delhi, India. 66 p.

Boyd, Claude E. 1971. Phosphorus dynamics in ponds. Proceedings 25th Ann. Conf. Southeastern Assoc. Game and Fish Commissioners: 418-426.

Boyd, Claude E., E.E. Prather, and Ronald W. Parks. 1975. Sudden mortality of a massive phytoplankton bloom. Weed Science. 23 (1): 61-67.

Clemens, Howard P. and Kermit E. Sneed. 1962. Bioassay and use of pituitary materials to spawn warm-water fishes. Research Report 61, Bureau of Sport Fisheries and Wildlife, United States Department of Agriculture (USDA). 30 p.

Crane, John S., *et al.* 1966. Togo fish project manual. United States Peace Corps. Oklahoma University, Norman, Olkahoma. 158 p.

Delmendo, Medina N. and Robert H. Gedney. 1974. Fish farming in pens - a new fishery business in Laguna de Bay. Laguna Lake Development Authority, Technical Paper 2. Pasig, Rizal, Philippines.

Denyoh, F.M.K. 1966. Pond fish culture development in Ghana. FAD World Symposium on Warm Water Pond Fish Culture. Rome, Italy.

Dillon, Olan W., Jr., *et al.* Warm water fish ponds. Farmer's Bulletin 2250. USDA. Washington, D.C. 14 p.

Dyche, L.L. 1914. Bulletin on ponds, pond fish, and pond fish culture. Part III. State Dept. of Fish and Game, Kansas. Kansas State Printing office, Topeka, Kansas. 130 p.

Eipper, A.N. and H.A. Gegier. 1965. Fish management in New York farm ponds. Cornell Extension. Bull. 1089. New York State College of Agriculture, Ithaca, New York. 39 p.

Fidler, Gary. 1973. Knowledge about your fish pond. Bureau of Fisheries and the United States Peace Corps, Manila, Philippines. 28 p.

Fijan, Nikola. 1966. Problems in carp fish pond fertilization. FAD World Symposium on Warm Water Pond Fish Culture. Rome, Italy.

Francis, Francis. 1865. Fish culture: a practical guide to the modern system of breeding and rearing fish. Routledge, Warne, and Routledge. London. 320 p.

Fridthjof, John. 1962. Encouraging the use of protein-rich foods. FAO, Rome, Italy. 103 p.

Gaines, John L , Jr., and Wilmer A. Rogers. 1975. Some skin lesions of fishes. The Pathology of Fishes. University of Wisconsin Press. Madison, Wisconsin: 429-441.

Gracia, Demetrio M. and Pio D. Bersamin. What you should know about carp culture. Philippine Fisheries Commission, Intramuros, Manila, Philippines. 7 p.

Gray, D. Leroy. 1970. The biology of channel catfish production. Agricultural Extension Service, Circular 535. University of Arkansas. 16 p.

Grizzell, Roy A., Jr., Olan W. Dillon, Jr., and Edward G. Sullivan. 1969. Catfish farming - a new farm crop. Farmer's Bulletin 2244. USDA. 22 p.

Hara, Shiro. 1972. Experiment on induced spawning of catfish (hito) stripping method and observations on the feeding of fry. P.F.C. Freshwater Fisheries Investigation Unit, Los Banos, Laguna, Philippines. 11p.

Hickling, C.F. 1961. Tropical inland fisheries. Longmans, Ltd. London. 287 p.

Hickling, C.F. 1968. The farming of fish. Pergamon Press, Ltd. London. 88 p.

Hickling, C.F. 1971. Fish culture. 2nd ed. Faber and Faber, London. 317 p.

Hora, S.L. and T.V.R. Pillay. 1962. Handbook on Fish culture in the Indo-Pacific region. FAD Fisheries Biology Technical Report 14. Rome, Italy. 204 p. (xeroxed copy)

Huet, Marcel, in collaboration with J.A. Tirmermans. 1970. Textbook of fish culture. Fishery News (Books) Ltd., London. 436 p. (translated from French by Henry Kahn)

Hutchinson, G. Evelyn. 1957. A treatise on limnology. John Wiley & Sons, Inc., New York. 1015 p.

Jeffrey, Norris B. 1969. Some aspects of the ecology of fish ponds. Proceedings 1969 Fish Faming Conf., Texas Agric. Extension

Service, Dept. Wildl. Science, College of Agriculture. Texas A & M University: 40-42.

Lagler, Karl F., John E. Bardach, and Robert R. Miller. 1962. Ichthyolgy. John Wiley & Sons, Inc., New York. 545 P.

Lawrence, J.M. 1949. Construction of farm fish ponds. Circular 92. Agric. Exp. Station, Auburn, Alabama. 55 p.

Lichtkoppler, Frank. Basic village pond fish production. U.S. Peace Corps, Madhya Pradesh, India. 11 p.

Maar, A., M.A.E. Mortimer, and I. Van der Lingen. 1966. Fish culture in central east Africa. FAO, Rome, Italy. 158 p.

Manual on Fishermen's Cooperatives. FAO Fisheries Studies 13. FAO, Rome, Italy. 124 p.

McLarney, William O. (ed.). 1973. The backyard fish farm workbook for 1973. Organic Gardening and Farming. Rodale Press Inc. The New Alchemy Institute, Woods Hole, Mass.

McLarney, William O. and J.R. Hunter. 1975. A new low-cost method of sealing pond bottoms. The Journal of the New Alchemists. 3: 85.

Meschkat, A. 1966. The status of warm-water fish culture in Africa. FAO World Symposium on Warm Water Pond Fish Culture. Rome, Italy.

Meyer, Fred P. Treatment tips - how to determine quantities for chemical treatments in fish farming. Bureau of Sport Fisheries and Wildlife. Fish Farming Exper. Sta., Stuttgart, Arkansas. US Dept. of the Interior. 20 p.

Meyer, Fred P., K.E. Sneed, and P.T. Eschmeyer. (eds.). 1973. Second report to the fish farmers. Resource Pub. 113. Bu. Sport Fish. and Wildl., USDI. 123 p.

Odum, Eugene P. 1971. Fundamentals of Ecology. 3rd ed. W.B. Saunders Co., and Toppan Co., Ltd., Tokyo, Japan. 574 p.

Ong, Kee Bian. 1968. Fish culture. Borneo Literature Bureau. Asiatic Lithographic Printing Press, Ltd. Hong Kong. 80 p.

Patino R., Anibal. Cultivo experimental de peces en estanques. Cespedesia II (5): 75-127. (translated by Wm. O. McLarney in the Journal of the New Alchemists. 3:86-90)

Prowse, G.A. 1968. Some basic concepts on fish culture. FAO Indo-Pacific Research Council, 13th Session. Brisbane, Queensland, Australia.

Rawson, G.C. 1966. A short guide to fish preservation. FAO, Rome, Italy. 67 p.

Report to the fish farmers. 1970. Resource Pub. 83. Bu. of Sport Fish. and Wildl., USDI. 124 p.

Rogers, Wilmer A. and John L. Gaines. 1975. Lesion of protozoan diseases in fish. The Pathology of Fishes. University of Wisconsin Press, Madison, Wisconsin: 117-141.

Samaka Service Center. 1962. The Samaka guide to homesite farminq. Samaka Service Center, Manila Philippines. 166 p.

Shell, E.W. 1966. Monosex culture of male *Tilapia nilotica* (Linn.) in ponds stocked at 3 rates. FAO World Symposium on Warm Water Pond Fish Culture. Rome, Italy.

Shook, Marilyn. 1974. Research Status Report: Experimental fish pen project. Bureau of Fisheries, Manila, Philippines. 7 p.

Sidthmunka, A., J. Sanglert, and O. Pawapootanon. The culture of catfish (*Clarias* spp.) in Thailand. Fisheries Dept., Bangkok, Thailand.

Swingle, H.S. 1957. Relationship of pH of pond waters to their suitability for fish culture. 9th Pacific Science Congr., Bangkok, Thailand.

Swingle, H.S. 1960. Comparative evaluation of two tilapias as pond fishes in Alabama. Transac. Am. Fish. Soc. 89(2): 142-148.

Swingle, H.S. 1966. Biological means of increasing productivity in ponds. FAO World Symposium on Warm Water Pond Fish Culture. Rome, Italy.

Swingle, H.S. 1966. Fish kills causes by phytoplankton blooms and their prevention. FAO World Symposium on Warm Water Pond Fish Culture. Rome, Italy.

Swingle, H.S., E.E. Prather, and J.M. Lawrence. 1953. Partial poisoning of overcrowded fish populations. Circ. 113. Agric. Exp. Sta., Auburn, Alabama. 15 p.

Swingle, H.S., B.C. Gooch, and H.R. Rabanal. 1963. Phosphate fertilization of ponds. Proceedings 17th Ann. Conf., South-eastern Assoc. Game and Fish Commissioners, Arkansas: 213-217.

Taverner, John. 1600. Certaine experiments concerning fish and fruite. London. 38 P. (reprinted 1968. Da Capo Press and Theatrum Orbis Terrarum Ltd., Amsterdam and New York).

Torrans, Eugene Leslie. 1973. Fish culture in Cameroon. Peace Corps Program and Training Journal. ACTION, Washington, D.C. 1(5): 14-47.

University of Rhode Island Marine Memorandum 30. 1972. Fisheries cooperatives: Their formation and operation. Marine Advisory Service. University of Rhode Island, Narragansett, Rhode Island. 18 p.

World Neighbors in Action. Raising fish in local farm ponds means protein and profit in Paraguay. World Neighbors International Headquarters, Oklahoma Cit, Oklahoma. 5(2-E).

Volunteers in Technical Assistance. 1975. Village Technology Handbook. VITA, Mt. Rainier, Maryland. 387 p.

Yashouv, A. Interaction between the common carp (*Cyprinus carpio*) and the silver carp (*Hyponhthalmichthys molitrix*) in fish ponds. Fish Culture Research Station, Dor, Israel.

MEASUREMENTS USED IN THIS MANUAL

1 gram (gm)	= 1000 milligrams (mg)
1 kilogram	= 1000 gm = 2.2 pounds (lb)
1 mg/l	= 1 part per million (ppm)
1 liter (l)	= 1000 milliliters (ml) = 0.26 gallons (gals)
1 inch (in)	= 2.54 centimeters (cm)
1 foot (ft)	= 30.5 cm
1 meter (m)	= 100 cm=1000 millimeters (mm) = 39, 37 inches
1 are	= 100 square meters (m^2)
1 hectare (ha)	= 10,000 m2 = 100 ares = 2.5 acres

° Centigrade (C) = 5/9 x (°F - 32)

° Fahrenheit (F) = (9/5 x °C) + 32